Harm de Blij's Geography Book

HARM DE BLIJ'S GEOGRAPHY BOOK

A Leading Geographer's Fresh Look at Our Changing World

Harm de Blij, Ph.D.

John Wiley & Sons, Inc.

New York • Chichester • Brisbane • Toronto • Singapore

In loving memory of
Gus
1977–1994

ACKNOWLEDGMENTS

It is a joy to know that thousands of elementary and high school students are again having the experience I had many years ago: being introduced to geography by an enthusiastic teacher. Mine was unforgettable: he not only taught me about distant lands and cultures, but he showed me how to do original research, how to keep notes, and how to write reports supported by maps and other illustrations. Eric de Wilde had lived in what was then the Dutch East Indies and in Western Europe, and at the Baarns Lyceum in the Netherlands he showed me how geography opened doors to wider worlds. At the time I thought it nothing unusual, but on many a day, after classes, a large group of us descended on the small town library to find out more about the countries and cultures he had described. Imagine it: we headed for the library rather than the ball field, on our own volition. He was a genius, he changed my life, and I have tried to emulate him ever since. I hope that this book reflects in a small way the measure of my gratitude to him—and to those who are embarked on a similar mission today.

During the past six years I have had the opportunity to share my enthusiasm for geography with a national television audience, and I thank Jack Reilly and Bob Reichblum for their support, and Jane Bollinger for her advocacy on behalf of geography's program segments. I also thank the several thousand viewers who have written to me over those years; their comments and questions induced me to write this book.

I acknowledge appreciatively the work of several people at my publisher. Carole Hall, Editor-in-Chief in the Professional and Trade Division, was in charge of this project throughout. Chris Jackson was editorial assistant, and Marcia Samuels managed the production. I also thank Stella Kupferberg for her interest, as well as Margie Schustack, Jennifer Holiday, Kara Raezer, Dee Dee DeBartlo, Lauri Sayde, and Kim Hendrickson. The cover design is Howard Grossman's.

Special thanks go to David and Janice Roland of The Roland Company in Reston, Virginia, without whose expertise and dedication this book would have not materialized. I am grateful also to Chuck Hyman and to Kevin Osborn, who designed this attractive layout; to Don Larson of Mapping Specialists in Madison, Wisconsin, for his help with the cartography; to Blair Burns Potter, managing editor of *Science News*, who assisted with copyediting; and to Robert Elwood, who compiled the index.

I am especially grateful to my colleague, Alan C. G. Best, who shares my enthusiasm for geography and who read the manuscript critically and productively.

My deepest appreciation goes to my wife, Bonnie, who was with me on many of the excursions described in these pages and whose unfailing good judgment and wise counsel were indispensable.

Harm de Blij
Boca Grande, Florida
January 1995

CONTENTS

PREFACE

The geographic perspective can throw new and useful light on so many subjects that no short book can cover them all. So I have made a selection based on a highly subjective survey: the mail resulting from my appearances as Geography Editor on ABC-TV's *Good Morning America* (GMA).

My first contact with ABC-TV occurred in April 1989, when the network, in search of a geographer to interview about the public's lack of geographic knowledge, called the University of Miami; the call was relayed to me in Baltimore where the annual meeting of the Association of American Geographers was in progress. I was asked to telephone one of GMA's producers, who asked me what I would say about geography "in a couple of minutes or so." I enthusiastically launched into my monologue only to be interrupted after 30 seconds. "No, no, that's not what we had in mind at all," she said. "We'd like you to stand in front of a map and point to some famous but out-of-the-way places and say a few words about them. You know, like Timbuktu and Katmandu." I suggested that for this, GMA wouldn't really need a professional geographer. "How about a high school student?" I suggested impoliticly. "We'll call you back," said the producer. I didn't count on it.

But GMA did call back, in the person of Jack Reilly, the show's executive producer. He asked me to stop by the office when next I was in New York. As it happened, I was to attend a dinner of the Explorers' Club just a couple of weeks later, and Reilly gave me an appointment.

I arrived for that appointment in doubt and left in awe. Like most television viewers I suppose, I had no idea what went on behind the scenes of a daily program such as GMA, nor of the energies, capacities, knowledge, insights, and other skills of the staff ranging from the on-air "talent" to the people on the studio floor. Jack Reilly had a keen sense of the effect on society of what he called "creeping geographic illiteracy." He as well as others on the GMA staff were well prepared for our discussion, having acquired a small geographic library, much of which they had obviously read. I argued that one brief segment during one show would do little to achieve Reilly's stated goal, which was to contribute toward an improvement of the situation by highlighting geography on the program.

While I left impressed by what I had learned about GMA, I wasn't sure I had made my case effectively. But a few weeks later I was asked to come back to New York for a kind of audition. This time I had nearly 15 minutes before the entire GMA staff to argue my point; the result was a five-day series to be aired during the autumn.

The series led to my appointment as Geography Editor on the GMA staff, effective in 1990. Since then, I've plied the dark and nearly empty streets of Manhattan on many an early morning on my way to the studio on 67th Street, and broadcast from remote places, including Scotland, Tasmania, Hong Kong, and Japan.

And I've learned about television's power. Some topics obviously leave viewers cold, but others generate some hot reactions. I get mail, lots of it, and so do others. People write to GMA, to two of my former universities (Miami and Georgetown), and even to professional geographic organizations of which I am a member. I learn a great deal from this correspondence, and the major topics of this book were chosen on this basis. If people are willing to take the time to sit down and write about something they saw on television, it surely must matter to them.

This book, therefore, consists of a series of extended talks on various topics, many of which were highlighted on *Good Morning America* and chosen by viewers through their reactions to them. Subjects range from those of immediate practical concern—using maps, coping with changing weather, understanding the new city—to broader issues such as the functions of political boundaries, the images and roles of capital cities, and the scramble for the Earth's last frontier, the oceans. The *Geography Book* is intended to help you navigate our challenging and changing world, and to constitute a guide to what lies ahead.

HARM DE BLIJ'S
GEOGRAPHY BOOK

C H A P T E R   1

GEOGRAPHY MEANS WIDER WORLDS

*Our world and our country change every hour of every day. Global population mushrooms.
The national debt grows. Old countries fall apart and new countries form. Familiar names
disappear from the map and new ones appear. Boundaries are created, erased, relocated.
Nations quarrel, even go to war over remote specks of land. The open oceans are becoming
less so. New terms come into use: Pacific Rim, ethnic cleansing, no-fly zone, choke point.*

*Like the human world, the physical Earth is being transformed. A "blizzard of the cen-
tury" is preceded by a hundred-year storm and followed by the worst flooding on record. New
temperature records seem to be set on a weekly basis. Unprecedented droughts afflict one
region while another drowns. And again newly prominent terms abound: greenhouse warming, El Niño, ozone layer, desertification.*

*Is there a common denominator for all this change? Can our world and its transforma-
tions be better appreciated through a particular perspective? This book answers both questions with one affirmation: geography.*

*I have a friend who says geography was the best cure for insomnia he ever found—but now
geography, too, has been transformed. Satellites observe the Earth and send information
unimagined just decades ago. Computers draw maps. Interactive atlases allow users not only
to look up places but also to see pictures, listen to local languages, hear traditional music. As a
result, geographers are much better at interpreting our changing world and at providing ideas
and information that can be the key to coping with it all.*

My love for geography (and such it truly is) stems from my very first experience with it in Nazi-occupied Holland during the Second World War. The autumn and winter of 1944–45 were a nightmare of danger and hunger. My parents were engaged in a daily battle for survival, and our village was a cold, gray, depressing place. At an age when I should have been out playing soccer and skating, I spent most of my time in the only slightly warmed room in our house, a bundle of clothes ready in case of emergency. But in that room were the books of a Dutch geographer named Hendrik Willen van Loon. Those geography books described worlds far away, where it was warm, where skies were blue and palm trees swayed in soft breezes. There were exciting descriptions of volcanoes and tropical storms, of maritime journeys to remote islands, of great, bustling cities, of powerful kingdoms and unfamiliar customs. I followed van Loon's journeys on the maps in an old atlas and dreamed of the day when I would see his worlds for myself. Van Loon's geographies gave me hope and, almost literally, a lease on life.

After the end of the war, my fortunes changed in more than one way. When the schools opened again, my geography teacher was an inspiring taskmaster who made sure that we, a classroom full of youngsters with a wartime gap in our early education, learned that while geography could widen our horizons, it also required some rigorous studying. The rewards, he rightly predicted, were immeasurable.

If, therefore, I write of geography with enthusiasm and in the belief that it can make life easier and more meaningful in this complex and changing world, it is because of a lifetime of discovery and fascination.

WHAT IS GEOGRAPHY?

As a geographer, I've often envied my colleagues in such fields as history, geology, and biology. It must be wonderful to work in a discipline so well defined by its name and so accurately perceived by the general public. Actually, the public's perception may not be so accurate, but people *think* they know what historians, geologists, and biologists do.

We geographers are used to it. Sit down next to someone in an airplane or in a waiting room somewhere, get involved in a conversation, and that someone is bound to ask: geography? You're a geographer? What *is* geography, anyway?

In truth, we geographers don't have a single, snappy answer. A couple of millennia ago, geography essentially was about discovery. A Greek geographer named Eratosthenes moved geographic knowledge forward by leaps and bounds; by measuring Sun angles, he not only concluded that the Earth was round but came amazingly close to the correct figure for its circumference. Several centuries later, geography was propelled by exploration and cartography, a period that came to a close, more or less, with the adventures and monumental writings of Alexander von Humboldt, the German naturalist-geographer. A few decades ago, geography still was an organizing, descriptive discipline whose students were expected to know far more capes and bays than were really necessary. Today geography is in a new technological age, with satellites transmitting information to computers whose maps are used for analysis and decision making.

Despite these new developments, however, geography does have some traditions. The first, and in many ways the most important, is that geography deals with the natural as well as the human world. It is, therefore, not just a "social" science. Geographers do research on glaciations and coastlines, on desert dunes and limestone caves, on weather and climate, even on plants and animals. We also study human activities, from city planning to boundary making, from wine-growing to churchgoing. To me, that's the best part of geography: there's almost nothing in this wide, wonderful world of ours that can't be studied geographically.

This means, of course, that geographers are especially well placed to evaluate the complicated relationships between human communities and societies, on the one hand, and natural environments, on the other; this is the second geographic tradition. Some of my colleagues investigate people's reactions to environmental hazards: why do people persist in living on the slopes of active volcanoes and in the floodplains of flood-prone rivers? How much do home buyers in California, for example, know about the earthquake risk on the spot where they buy? (And what are they told by real estate agents?) And there's another aspect to this environmental field: people's health. Yes, there are geographers who study illnesses geographically. Where are diseases found? How do they spread? Are there specific kinds of environments in which certain illnesses are located? How are disease and environment related? How do the changes that societies undergo affect the health of their populations? Where do people go when they seek health care? How does the location of a health-care facility affect the opportunity to obtain care? Some of these medical geographers have made huge contributions to the betterment of humanity.

A third geographic tradition is simply this: we like to know and understand different cultures and distant regions. In the past, it was rare to find a geographer who did not have some considerable expertise in some foreign area, large or small. Most spoke one or more foreign languages, read the newspapers from that part of the world, studied in the field, and did research there. That tradition has faded a bit in the new age of satellite data and computer cartography, but many students still are first attracted to geography because it aroused their curiosity about some foreign place.

And finally, there is what we like to call the "location" tradition. Why are things located on the Earth's surface where they are? What does their location mean to their prospects? You can imagine the wide range of applications of this question. Why did one city thrive, while a nearby settlement dwindled and failed? Often the geographic answer illuminates historic events. More currently, where should that new shopping center be built? If there are, say, three options, its geography will be crucial. Many geographers get into the planning business.

Still we don't have a short-answer definition of geography.

LOOKING AT THE WORLD SPATIALLY

To pull it all together, we need a word that telegraphs our main geographic preoccupation, and that word derives from space—not celestial space, but earthly space. We geographers look at the world spatially. I sometimes try this concept on a questioner: historians look at the world temporally or chronologically; we geographers look at it spatially. With a little luck, my neighbor will nod wisely, take out his *USA Today*, and read about the latest geographic illiteracy test.

But for some people, even geographers, one word is not enough. When the National Geographic Society decided to help revive school geography, it appointed a small committee led by a noted geographic education specialist, Christopher "Kit" Salter, to develop a set of geographic themes that would form the basis of the campaign. In 1986, the Society published the result: several million copies of an annotated map entitled *Maps, the Landscape, and Fundamental Themes in Geography*.

Those themes are now thoroughly entrenched in geographic education and reflect geography's spatial preoccupation: three themes focus on interactions between people and environments, on regions of the world, and on questions of location; two focus on place and movement. The latter two could actually be subsumed under the first three, and there was a good deal of raucous argument about this during the project's development. But the five-theme notion prevailed.

You get the picture: it's impossible even for geographers to come up with a generally accepted definition of their discipline. Actually, this is one of geography's strengths. Under the "spatial" umbrella, we study and analyze processes, systems, behaviors, and countless other topics that have spatial expression. It's the tie that binds geographers, this interest in patterns, distributions, diffusions, apportionments, allocations, circulations—the ways in which the physical and human worlds are laid out, interconnect, interact, and transform. But under this umbrella, too, we can go our own way, to pursue our own special interests. I have colleagues who do research on sports (Where do football players come from? What factors determine where they go to college and, in some cases, whether they go from college to the pros?), on wine (Where are the best climatic areas for winegrowing outside of France?), on religion (How are patterns of church membership changing?), on health (What are the diffusion routes of AIDS?), on climate (How valid are greenhouse-warming predictions?) and on literally countless other topics. I'm always fascinated to hear what they're discovering, and as I tell my students, the Age of Discovery may be over, but the era of geographic discovery never will be.

THE SPATIAL SPECIALIZERS

Geographers take not only a wide view, but also a long view. We try not to lose sight of the forest for the trees, to put what we discover in temporal as well as spatial perspective. "Geography is synthesis" is one popular answer to that question about just what geography is. As you will see later in this book, some daring generalizations can set research off in very fruitful directions. These days, though, it takes courage to generalize. This, as we all know, is the age of specialization.

And yes, geographers specialize too. Some of them, as we just noted, specialize in some region of the world. Government agencies find such area specialists useful as consultants, and increasingly, we find, businesses recognize their value.

Others specialize in some field of geography rather than a region. Many geographers, for example, study various spatial aspects of cities and other urban places. These urban geographers look at the city from spatial standpoints, and their studies range from highly analytical research on land values and rents to speculative assessments of intercity competition. For example, when Germany was reunified, the question of its capital city arose: should it be Bonn, the old West German capital near the western borders of the country, or Berlin, with its wartime legacy, near the eastern (Polish) border? Some geographers, myself included, saw merit in a third choice, a new capital in a more central location with no "baggage" from times of war and division. One well-situated city that might have been considered was Hannover, but the decision to move Germany's capital to Berlin was made quickly, before such alternatives could be tested. Now, many Germans have second thoughts about it.

Other geographers combine economics and geography, and focus on spatial aspects of economic activities. The rise of the world's new economic giants on the Pacific Rim has kept them busy.

Still others focus on spatial aspects of political behavior. Political scientists tend to focus on institutions, political geographers on political mosaics. Geopolitics, an early subfield of political geography, was hijacked by Nazi ideologues and lost its reputation; but recently, geopolitics has been making a comeback as an arena of serious and objective research. From power relationships to boundary studies, political geography is a fascinating field.

There are literally dozens of fields of specialization in geography, and for students contemplating a career in geography it's a bit like being in a candy store. Interested in anthropology? Try cultural geography! Biology? There's biogeography! Geology? Don't forget geomorphology, the study of the evolution of landscape. Historical geography is an obviously fruitful alliance between related disciplines. The list of such options is long, and it is still growing. Developments in mapmaking have opened a whole new horizon for technically inclined geographers, as we will see in the next chapter.

Over a lifetime of geographic endeavor, many geographers change specialties, and I'm one of them. I was educated to be a physical geographer, that is, as someone who specializes in landscape study (geomorphology) and related fields. As such, I spent a year in the field in Swaziland, in southern Africa, trying to determine whether a large, wide valley there was part of the great African Rift Valley system, the likely geographic source of humanity. While I was preparing for this research, however, I met a political geographer named Arthur Moodie, a British scholar who came to Northwestern University as a visiting professor. I took his classes and never forgot them. When I was hired by Michigan State University as a physical geographer, I also continued to read and study political geography. Eventually, I was asked to teach a course in that field, wrote a book and some articles about it, and thus developed a second specialization.

What I didn't realize, at first, was how my background in physical geography would make me a better political geographer. Like geopolitics, environmental determinism had acquired a bad name between the World Wars, and it could be dangerous, professionally, to try to explain political or other social developments as influenced by environmental circumstances. But I knew that, like geopolitics, environmental studies would make a comeback. When they did, I had the background to participate in the debate. That's how, many years later, I was appointed to Georgetown University to teach environmental issues in the School of Foreign Service.

I made only one other foray into a new field, and that was also as pleasant a geographic experience as I've ever had. It all began with a great bottle of wine. A fateful dinner with that bottle of 1955 Chateau Beychevelle so aroused my curiosity that, five years later, I was working on my book entitled *Wine: a Geographic Appreciation*, was teaching a course called The Geography of Wine at the University of Miami, and saw some of my students enter the wine business armed with a background they often found to be very advantageous. Geography has few limits—and specialization does have its merits.

BUT IS GEOGRAPHY IMPORTANT?

Remember the bumper sticker, popular some years ago, that said "If You Can Read This, Thank a Teacher"? One day I was driving down one of my least favorite highways, Interstate 95 between Fort Lauderdale and Miami in Florida, when a car passed me whose owner had modified that sticker by inserting the word "Map" after "This" and by pasting a piece of road map at the end of the slogan. I didn't need to ask what that owner's profession was. A geography teacher, obviously.

The fact is, a lot of us cannot read maps. Surveys show that huge numbers of otherwise educated people don't know how to use a map effectively. Even simple road maps are beyond many more of us than you might imagine. People who, you would think, deal with maps all the time and therefore know how to get the most out of them—travel professionals—often have trouble with maps. Let me give you just one example from my network experience.

In April 1989, I had been asked to help prepare a five-part series on geography to be aired in September. During the summer, working with my producer, the late Sonya Selby-Wright, I wrote and rewrote hundreds of pages of script. Finally, in August, we were ready to tape. Our first in-the-field taping would take place in the vicinity of Traverse City, in the northern part of Michigan's Lower Peninsula. ABC's travel office made all the arrangements for the camera crew and field producer; I flew in from Florida, where I was based at the time, and met my former colleague from Michigan State University, Harold "Duke" Winters, in Lansing on the way north. He would appear with me on this program discussing the glacial history of the Traverse City area.

It was still summer, and the days were long. Winters and I arrived at Traverse City in late afternoon, and we had dinner. The producer arrived from Chicago, where she had been working, at about 8 P.M. But there was no sign of the camera crew, not at 9, or 10, or 11 P.M. Since we were due to convene at dawn, it looked as though our shoot was in trouble.

Finally, just before midnight, when Winters and I stood on the hotel's veranda wondering whether there would in fact *be* a crew for us, two Lincoln Continentals roared into the driveway, loaded with crates, boxes, tripods, cameras—and three angry people. They had been on the road for six hours or so, and, as one of them put it, "every time we looked at the map, Traverse City seemed farther away than it was before." What had happened? The people who organized their trip had sent them to Michigan alright, but to Detroit! After all, Michigan was Michigan; just rent a couple of cars and drive over to Traverse City. They started out on the road in the late afternoon, looking forward to a nice dinner in this resort town, and then, when it was too late to do anything about it, they realized that the people back in New York had little idea that driving from Detroit to Traverse City isn't like driving from La Guardia to Poughkeepsie.

When I told Jack Reilly, the executive producer, the story a few weeks later he just shook his head. "I guess it's worse than I thought," he said, referring to an earlier chat we had had about a poll that showed 47 percent of respondents unable to mark New York State on an outline map of the 48 states.

Okay, you may say. As an everyday tool to make life a bit more predictable and efficient, geography has its uses. But does that make it important?

Consider this: a general public not exposed to a good grounding in geography can be duped into believing all kinds of misinformation. Even today, despite the best efforts of the National Geographic Society and its allies, an American student might go from kindergarten through graduate school without ever taking a single course in geography—let alone a fairly complete program. (That's not true in any other developed country, nor in most developing ones. Geography's status is quite different in Britain, Germany, France, and such countries as Brazil, Nigeria, and India.) This means that when a group of scientists decides to scare the beejeebers out of the public by predicting imminent glaciation (as they did in the 1960s) or looming greenhouse warming (the fad of the 1990s), far too many people are insufficiently skeptical and, through their elected representatives, may be persuaded to spend billions better invested elsewhere.

Equally important is geography's function as an antidote to isolationism. Can there be a more crucial objective than this? In our ever more interconnected, overpopulated, competitive, and dangerous world, the more we know about our planet and its environmental history and geography, about other peoples, cultures, political systems, and economies, about precarious boundaries and sensitive borderlands, about distant resources and developing regions, the better prepared we will be for the challenging times ahead.

From that perspective, geography's importance is second to none.

DISCOVERING GEOGRAPHY

More and more people are discovering how interesting, challenging, even important geography can be. Actually, they're rediscovering something Americans once knew very well. During the half century following the First World War, thus including the Second World War, geography was a prominent component of American education. In prewar debates, wartime strategy, and postwar reconstruction, geographers played useful, sometimes crucial roles. Geographers were the first to bring environmental issues to public attention. They knew about foreign cultures and economies. They had experience with the workings of political boundaries. They produced the maps that helped guide United States policies.

In the 1950s and early 1960s, Americans continued to be well versed in geography. American success during the Second World War had drawn our attention to the outside world as perhaps never before. Maps, atlases, and globes sold by the millions. The magazine with geography's name on it, *National Geographic*, saw its subscriptions grow to unprecedented numbers. University geography departments enrolled more students than they could handle. When President John F. Kennedy launched the Peace Corps, geographers and geography students were quickly appointed as trainers and staffers.

But, as so often happens when social engineers get hold of a system that's working well, the wheels came off. Professional educators thought they had a better idea about how to teach geography: rather than educating students in disciplines such as history, government, and geography, they would teach these subjects in combination. That combination was called social studies. The grand design envisioned a mixture that would give students a well-rounded schooling, a kind of civics for the masses, which implied that school teachers would no longer be educated in the disciplines either. They, too, would study social studies.

Prospective teachers from the School of Education had been among my best and most interested students at the University of Miami during the early 1970s. They registered in large numbers in two courses: World Regional Geography, which was an overview of the geography of the wider world, and Environmental Conservation, a course that was years ahead of its time, and to which even the Department of Biology sent its students. But when the social studies agenda took effect, the student teachers stopped coming. They now had other requirements that precluded their registration in geography.

We geographers knew what this would mean and what it would eventually cost the country. The use of, and knowledge of maps would dwindle. Environmental awareness would decline. Our international outlook would erode. Our businesspeople, politicians, and others would find themselves at a disadvantage in a rapidly shrinking, ever more interconnected—and competitive—world. Many of us wrote anguished letters to government agencies and elected representatives, to school district leaders and school principals. Fortunately, many private and parochial schools continued to teach geography. But for public education, the die was cast.

REVERSAL OF FORTUNE

It took just 10 years to create exactly what geographers had predicted: a kind of geographic illiteracy. Of course, all of us had already experienced it in our college classrooms. I well remember giving a lecture on the competition for land between villagers and wildlife in Kenya, complete with a large map of East Africa and pictures of acacia trees and elephants, when, at the end of the class, a student asked on what continent all this was happening. He associated elephants with India.

Such incidents would not have attracted national attention any more than our letters had, were it not for an accidental sequence of events. The first event involved a colleague of mine at the University of Miami who liked to start his class by asking students to identify a number of prominent geographic locations on a blank map of the world's countries. The results were always abysmal, and they grew worse as time went on. The good professor would grade the class as a whole and, reportedly with biting sarcasm, would announce the large percentage of participants who could not locate the Pacific Ocean, the Sahara, Mexico, or China.

Early in the fall semester of 1980, the student newspaper, the *Miami Hurricane*, got hold of the test, a summary of the test results, and the professor's witty commentary. The paper's front-page story on this tale of "geographic illiteracy" was picked up by the major news media. NBC's *Today* show appeared on campus. ABC's *Good Morning America* invited the principals to New York, but the segment was too brief to throw real light on the problem.

The news, however, had spread throughout the country, and while officials at the University of Miami fretted about what the story might do to the university's reputation, teachers elsewhere tried their own tests on their students. We are all too familiar with the results. At one Midwestern college, only 5 percent of the students could identify Vietnam on a world map. At another college, only 42 percent correctly named Mexico as our southern neighbor. Specialists, including some of the educators who had engineered the demise of school geography, claimed to be "dismayed" at such results. Geographers were not surprised, but the question was: how do we reverse this ignominious tide?

RECOVERY

Help came from the most unexpected quarters. An American President, upon arriving in the capital of Brazil, pronounced himself pleased to be in Bolivia. (Later, an American Vice President, at the top of the stairs of Air Force Two in a Pacific country, apparently forgot the name of the place and addressed his hosts as officials of the "land of happy campers.") The National Geographic Society commissioned a Gallup survey that proved without a doubt that American students had fallen far behind in their geographic knowledge. Newspaper and television reports of these and other manifestations of our collective geo-blindness led to calls for action: resurrect geography! Extract it from the social studies! Bring it back to the schools!

All this might have happened in due course, but the intervention of the National Geographic Society speeded up the process.

To most observers, it would have seemed natural for an organization known as the National *Geographic* Society to come to the aid of the discipline. But it was not so simple. For many years, the Society and the discipline had not enjoyed good relations. To the Society and its leadership, professional geographers seemed snobbish, insulated, and often unimaginative. To professional geographers, the Society's popularization of its magazine under the rubric of geography was inappropriate and misleading. "There's precious little geography in the *National Geographic*," said my professor at Northwestern University when I arrived there as a graduate student in 1956. "If you're going to subscribe, you'd better have the magazine sent to your home. Not a good idea to see it in your departmental mailbox here."

That amazed me. In fact, when I was living in Africa during the early 1950s, the *National Geographic* was my window to the world, its maps a source of inspiration. I had written its president, Gilbert H. Grosvenor, in 1950 to tell him so. He sent a gracious letter in response, urging me to continue my interest in geography and inviting me to visit the Society's headquarters "if [I] were ever to come to the United States." But as a graduate student, I soon realized that the National Geographic Society and its publications were not held in high esteem among "professional" geographers generally.

Against this background, my hopes for help from the Society in the 1980s were not high, but because I had been elected to the Society's Committee for Research and Exploration, I was in a better position than almost anyone else to press the issue on its president, Gilbert H. Grosvenor's grandson. It was always my notion that assistance from the Society should go to colleges and universities, and to the professional organizations of the discipline, of which, as will be noted, there are several. But when the Society eventually put its shoulder to the geographic wheel, it was not to help college geography. Rather, the campaign, underwritten by tens of millions of dollars and a huge investment in staff and production, went to the schools and to teachers.

Of course, the Society was right.

THE REST IS GEOGRAPHY

The rest, as they say, is history—or rather, geography. The National Geographic Society formed a coalition with the National Council for Geographic Education, the Association of American Geographers, and the American Geographic Society to promote the teaching of geography in the schools. A nationwide, state-by-state alliance for geography teachers was funded by the Society. An annual National Geography Awareness Week was instituted, with the endorsement of the U.S. Congress. And an annual National Geography Bee, modeled on the famous spelling bee, was introduced. Largely through the Society's efforts, geography was established as one of the initial five cornerstones for American education during the tenure of Education Secretary Lamar Alexander in President George Bush's administration.

In 1984, I was appointed editor of the National Geographic Society's new scholarly journal, *National Geographic Research*, and established a second home in Washington, D.C. With an office in the then-new headquarters building at 1600 M Street, N.W., I was able to watch the geography effort at close range. Much of it was exhilarating, although there were those on the Society's staff who did not approve of the huge expenditures being made on behalf of geography. Some of it was disquieting, because the Society, with the help of a few compliant academics, set about not only in support of geography, but also in reform of it.

Geography, however, is once again in the public eye. Books about geography are popular. A PBS television series called *Where in the World Is Carmen Sandiego?* is making geography fun for kids. For the first time in many years, my colleagues and I are able to tell the general public, and not just our students, how exciting modern geography is, how much has changed in this field, and how useful a knowledge of it is in our changing world.

CHAPTER 2

MAPS: THE LANGUAGE OF GEOGRAPHY

Geography could not exist without maps. Maps are the language of geography. By means of maps, geographers communicate their ideas, pose their questions, solve their problems. Dull tables of statistical data come to life when they are represented as maps. Resource discoveries, economic trends, political changes, environmental variations, and countless other developments can be described in long, obtuse paragraphs of prose—or revealed by clear, vivid maps.

If a picture is worth a thousand words, then a map is worth a million. This is no mere generalization. The United States Geological Survey (USGS) publishes a set of Quadrangle sheets in the 7.5 Minute Series. An average sheet, it is estimated, carries a million items of information. Some of these maps cover more complex terrain than others, so there is some variation. But if you have not seen the USGS map for your area, even if it is an urban area, you're in for a pleasant surprise. These are not maps of the geology of your area, but of the surface: the slopes and streams, roads and paths, forests and lakes, towns and farms, and virtually all else in the natural and the cultural landscape. Like a good book, such a map is hard to put down once you have started "reading" it.

READING MAPS

There is more to reading a map than first meets the eye. To get the most out of a map we must be able to interpret everything it represents. This can be quite a challenge, because some maps carry many symbols. Since maps represent on a sheet of paper (or on a globe, or a television or computer screen) a part of the real

Facing Page: *Detail from a USGS Quadrangle sheet for the area around Mount Rainier in the State of Washington. Contour lines and other topographic markings are clearly illustrated.*

world, mapmakers use symbols to convey elements of that real world. Some of these are easy to make out: large dots for cities, smaller dots for towns, still smaller dots for villages. Highways often are represented by double lines, two-lane roads by single lines, and tracks or paths by dashed lines. But for other symbols, it is necessary to consult the legend of the map. Different maps represent certain features, such as houses, forested areas, pipelines, or power stations, in different ways. Only a careful check of the legend will ensure a correct interpretation of what the map shows.

One especially difficult challenge for cartographers, as mapmakers are often called, is the representation of the hills and valleys, slopes and flatlands of an area, collectively called its *topography*. This is done in various ways. One is to create an image of sunlight and shadow so that the wrinkles of the topography are alternately lit and shaded, creating a visual image of the terrain. Another, technically more accurate way is to draw contour lines, as is done on USGS Quadrangle sheets (opposite page). A contour line connects all points that lie at the same elevation. A round hill rising above a plain, therefore, would appear on the map as a set of concentric circles, largest near the base and smallest near the crest. If the contour lines are bunched closely together, the hill's slope is steep; if they lie farther apart, then the slope is gentler. Contour lines can represent scarps, hollows and all other features of the local topography. At a glance, we can tell whether the *relief* in the area (the vertical distance between its high and low points) is great or small: a "busy" contour map means lots of high relief.

DIRECTION

Foreign visitors to the United States tell me that one of their first impressions of Americans is how good our sense of direction can be. Europeans, many of whom come from the old, mazelike cities of their realm, are not used to instructions that say "Go four blocks east on Twenty-third Street and three blocks north on Fifth Avenue."

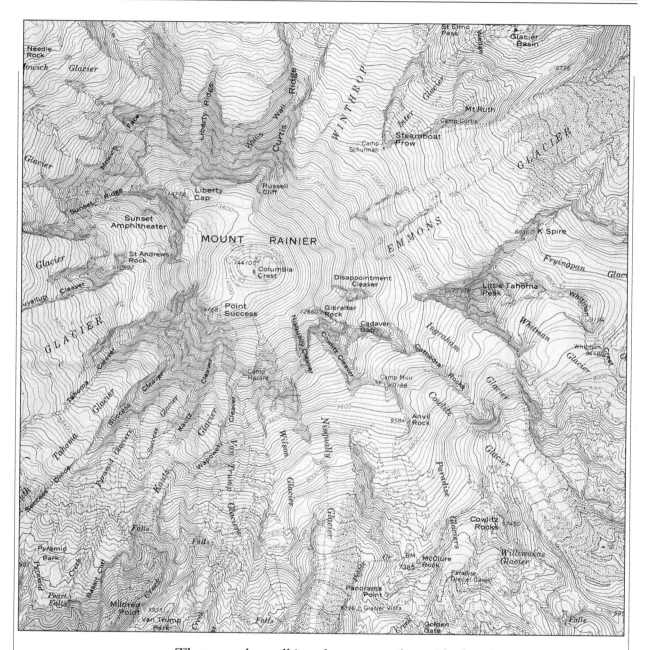

They may be walking the rectangular grid of an American city's downtown, but their awareness of what is north, south, east, or west is low. In most European cities, the compass directions simply aren't useful when it comes to finding the way. Americans, on the other hand, are likely to think first of the *compass rose*. Ask about someone's home suburb, and the first reaction often includes a directional reference: "Rosewood lies about six miles west of the city...."

But this does not mean that directions cannot be confusing, even to American map readers. It is convention that north lies at the top of a map, south at the bottom, east to the right, and west to the left. Yet some maps are not aligned this way, and even experienced map readers can be confused if this is not made clear by a prominent "north arrow" pointing in some direction other than upward. So in addition to checking the legend's symbols, a map reader needs to check the map's general orientation.

The compass rose consists of more than the four main directions. Midway between north and east the direction is northeast, and between north and northeast it is north-northeast; between northeast and east it is east-northeast. These refinements are of use mainly in navigation. Which leads us to a useful extension of the compass rose to our wrist watch. Imagine that you are standing on the deck of a boat, headed in a certain direction. That direction may be west-northwest or south-southeast. You spot a school of dolphins, just off to the right of your course, ahead of the bow. Rather than calling out the compass rose direction, a quicker directional reference would be "Dolphins at one o'clock!" Using the clock so that you are always moving toward twelve o'clock is a great way to share quick information on the highway. "Elk in the meadow at three o'clock!" will have all faces pointed in the proper direction; "Elk over on the left!" is far less specific. When I took students to Africa on safari, we always practiced this method—often with excellent results.

One final point involving direction. Why is it that the world is always represented in such a way that Europe and Asia and North America are at the top, and Australia and Antarctica are at the bottom? Again, this is a matter of convention. While it is logical to use the Earth's two poles as top and bottom of any world map, nothing in nature specifies that the North Pole should be at the top of the map and the South Pole at the bottom. What is now universal practice developed from the work of the earliest cartographers, who lived in the Northern Hemisphere and who started at the top of their page. Most of what was to be discovered turned out to lie to the east, west, and south of their abodes, and so Africa, South America, and Australia came to occupy the bottom half of the evolving world map. It has been that way ever since, except in Australia and New Zealand. There they pointedly draw maps that put "Down Under" on top of the world.

SCALE

A map represents a part or all of the real world on a small piece of paper. The larger the area represented, the smaller the *scale* of the map and the less detail it can display. Imagine it this way: on a page of this book, you can draw a fairly detailed map of the city block or suburban street (or village or farm) where you live. At such a large scale, you can show individual houses, streets, and sidewalks. Most of this detail would be lost, however, if the page were to contain a map of an entire city. Now the scale would be smaller, and only major urban areas and arteries could be shown. Put an entire state on the page, and the city becomes little more than an irregularly shaped patch. In turn, the state becomes just an outline if the page must contain the whole country, with very little detail possible.

Each time, in the example above, we made the map's scale smaller to accommodate an ever larger area on our page. Thus, in addition to the legend and its symbols, we should examine the scale of any map we read. Our expectations of what the map can tell us are based in part on the scale to which it is drawn.

Why is scale smaller when the area represented becomes larger? Because scale refers to a ratio: the ratio of a distance or area on a map to the actual, real-world distance or area it represents. To simplify, let us use the number one for the distance on the map. On our map of a city block or suburban street, 1 inch would represent about 200 feet, or 2,400 inches, so the scale would be 1:2,400. This ratio can also be represented as a fraction: 1/2,400. To get the whole city on

MEASURING TIME: A MATTER OF LINES

The measurement of time on the Earth's surface is important in the study of our planet. One of the most obvious ways to start dealing with time is to use the periods of light and darkness resulting from the daily rotation of the Earth. One rotation, one cycle of daylight and nighttime hours, constitutes one full day. The idea of dividing the day into 24 equal hours dates from the fourteenth century.

With each place keeping track of its own time by the Sun, this system worked well as long as human movements were confined to local areas. But by the sixteenth century, when sailing ships began to undertake transoceanic voyages, problems arose, because the Sun is always rising in one part of the Earth as it sets in another. In a sea voyage, such as that of Columbus in the *Santa Maria*, it was always relatively simple to establish the latitude of the ship. Columbus had only to find the angle of the Sun at its highest point during the day. Then, by knowing what day of the year it was, he could calculate his latitude from a set of previously prepared tables giving the angle of the Sun at any latitude on a particular day. However, it was impossible for him to estimate longitude. In order to do that, he would have to know accurately the difference between the time at some agreed meridian, such as the prime meridian (zero degrees longitude), and the time at the meridian where his ship was located. Until about 1750 no portable mechanical clock or chronometer was accurate enough to keep track of that time difference.

With the perfection of the chronometer in the late eighteenth century, the problem of timekeeping came under control. But by the 1870s, as long-distance railroads began to cross the United States, a new problem surfaced. Orderly train schedules could not be devised if each town and city operated according to local Sun time, and the need for a system of time organization among different regions became essential. The problem of having different times at different longitudes was finally resolved at an international conference held in 1884 in Washington, D.C. There, it was decided that all Earth time around the globe would be standardized against the time at the prime meridian, which passed through Britain's Royal Observatory at Greenwich. The Earth was divided into the 24 time zones, each

the page, 1 inch would have to represent 2 miles, or 126,720 inches. Now the ratio (1:126,720) becomes a much smaller fraction: 1/126,720. As maps go, 1:126,720 still is a pretty large-scale map. To get an entire medium-sized state on our page, our scale would have to drop to about 64 miles to the inch, or 1:4,000,000. And for the whole United States, the scale would be 1:40,000,000.

The scale of a map, therefore, tells us much about its intended use. When a developer lays out a new subdivision or a planner considers the placement of a new shopping center, large-scale maps are needed. The useful road maps made available by the American Automobile using the time at standard meridians located at intervals of 15 degrees of longitude with respect to the prime meridian (24 x 15 degrees = 360 degrees). Each time zone differs by one hour from the next, and the time within each zone can be related in one-hour units to the time at Greenwich. When the Sun rises at Greenwich, it has already risen in places east of the observatory. Thus, the time zones to the east are designated as *fast*; time zones west of Greenwich are called *slow*.

This solution led to a peculiar problem. At noon at Greenwich on January 2, 1996, it will be midnight on January 2 at 180 degrees east longitude (12 time zones ahead) and midnight on January 1 at 180 degrees west longitude (12 time zones behind). However, 180 degrees east and 180 degrees west *are the same line*. This line was named the *International Date Line* by the Washington conference. It was agreed that travelers crossing the date line in an eastward direction, toward the Americas, should repeat a calendar day; those traveling west across it, toward Asia and Australia, should skip a day. The date line did not pass through many land areas (it lies mainly in the middle of the Pacific Ocean),

thereby avoiding severe date problems for people living near it. Where the 180th meridian did cross land, the date line was arbitrarily shifted to pass only over ocean areas. Similarly, some flexibility is allowed in the boundaries of other time zones to allow for international borders, and even state borders in such countries as Australia and the United States. Some countries, such as India, choose to have standard times differing by half or a quarter of an hour from the major time zones. Others, such as China, insist that the entire country adhere to a single time zone.

A further arbitrary modification of time zones is the adoption in some areas of daylight saving time, whereby all clocks in a time zone are set forward by one hour from standard time for at least part of the year. The reason for this is that many human activities start well after sunrise and continue long after sunset, using considerable energy for lighting and heating. Energy can be conserved by setting the clocks ahead of the standard time. In the United States today, most states begin daylight saving time during the first weekend in April and end it on the last weekend in October.

Association and by state tourist offices are at medium scales. Maps of world distributions (of, say, population growth by country) can be presented at small scales. The map's function is the key to its scale.

HOW A MAP DEFEATED CHOLERA

Cholera epidemics were a scourge of the nineteenth century. In London during the 1850s, thousands died. No one knew the cause of the disease, and fear reigned.

Enter physician-geographer John Snow, who for many years studied cholera's repeated outbreaks. In 1854 he tried a new tack:

focusing on the hard-hit Soho district, he drew a large-scale map of the city blocks and marked with a dot the place of residence of every one who had died and every new victim.

The map, after many months, revealed a clear pattern: a majority of Soho's hundreds of deaths and thousands of cases concentrated around a prominent intersection along Broad Street. And there, in the middle of the street, stood the suspect—the communal pump from which everyone in the vicinity drew their water.

Dr. Snow took his map to the city's authorities, who at first were reluctant to inactivate the pump. But he persuaded them to act, and the handle was removed. Almost immediately, the area's new cases dropped to zero. The link between water and cholera had been established—by geography.

MAP PROJECTIONS

Even at the same scale, two maps of the world can look quite different. Take a look at Greenland. On some maps, Greenland is almost the same size as South America. But in reality, Greenland is less than *one-eighth* as large as the continent. How can this happen? The answer lies in the *projection* used by the cartographers to produce the map. By manipulating the map projection, you can exaggerate, diminish, distort, and otherwise modify any part of the Earth's surface. This means that maps can be used for propaganda purposes, or worse. Countries can be made to look larger, compared to other countries, than they really are. Places can be made to look closer than they really are. Such maps can produce fear, intimidation, aggression, or, at the very least, misjudgments among their readers. So, as with the written word, the rule is: reader beware!

For centuries, mapmakers have faced the key problem: how to represent our spherical Earth on a flat surface. To get the job done, they constructed an imaginary *grid* around the planet, using the poles of rotation and the globe-dividing equator as their starting points. Since a full circle is 360 degrees, the Earth was divided, pole to pole,

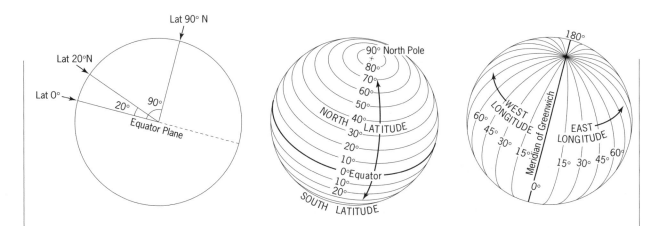

Modern globes use a grid system consisting of horizontal lines called parallels and vertical lines called meridians. All parallels north and south of the equator are designated by their angular position, such as 20° north latitude (Lat 20°N).

by *meridians*. Of course a starting meridian, or *prime meridian*, was needed, and this would be zero degrees. That decision was made when the British Empire was at its zenith, and so, not surprisingly, the prime meridian was established as the line of *longitude* running through the Greenwich Observatory near London. That turned out to be a fortunate choice, because the 180-degree line, which would divide the globe into Western and Eastern Hemispheres, lay on the opposite side of the world from London—right in the middle of the Pacific Ocean.

The meridians are the "vertical" lines of the Earth's grid; they converge on the poles and are farthest apart at the equator. The "horizontal" lines, called *parallels*, meet nowhere: they form equidistant rings around the globe, starting at the equator. These parallels, too, are numbered by degree: the equator, being neither north nor south, is zero degrees *latitude*. Then the numbers go up. Mexico City lies just below 20 degrees north latitude; Madrid, Spain, just above 40; St. Petersburg, Russia, at 60; and Russia's northern islands around 80. The North Pole, of course, lies at 90 degrees, the highest latitudinal point on Earth.

To identify what geographers call the *absolute* location of a place, we must state the degrees, minutes, and seconds north or south latitude. That seems cumbersome at first, but when you get used to it you can quickly find the remotest spot anywhere on the globe. Importantly, no two places have exactly the same location. Very recently, an amazing piece of technology has come into use: the Global Positioning

System (GPS), which operates via satellite command. You can hold it in your hand and it will display on its screen your exact location on the Earth's surface, in terms of latitude and longitude. The potential of GPS technology for mapmakers can only be imagined.

WHY ARE MAP GRIDS CALLED PROJECTIONS?

Having established their global grid, the cartographers now had a frame of reference to use in mapping the round Earth on a flat surface. Since all parallels and meridians intersect each other at right angles, it is possible to lay out a rectangular grid on that basis. This is what the Flemish mapmaker Gerardus Mercator did in 1569, thus inventing a layout that is still in use today. You can spot a Mercator map (opposite page) quickly: not only do the parallels and meridians intersect at right angles, but Greenland is nearly as large as South America. And Antarctica becomes a globe-girdling landmass, not a polar cap.

That's the problem with projections: they distort the real, ball-shaped world. Near the equator, Mercator's projection is fairly accurate, and you can use the scale printed on the map. But the farther north or south you go, the more exaggerated are the sizes of landmasses and water bodies. Europe, especially northern Europe, is much larger on a Mercator map than it should be.

You might think that such size discrepancies would have sent Mercator's projection into oblivion, but that did not happen, for two reasons. First, for all its size distortion, Mercator's layout did have one major virtue: all directions remain accurate. For navigators, therefore, Mercator's map was a lifesaver. And, truth be told, leaders and teachers in midlatitude countries liked a map that showed their countries relatively large.

For nearly a century, the National Geographic Society often used the Mercator projection to display world political changes. Then, during the 1980s, the Society's directors decided to adopt a different

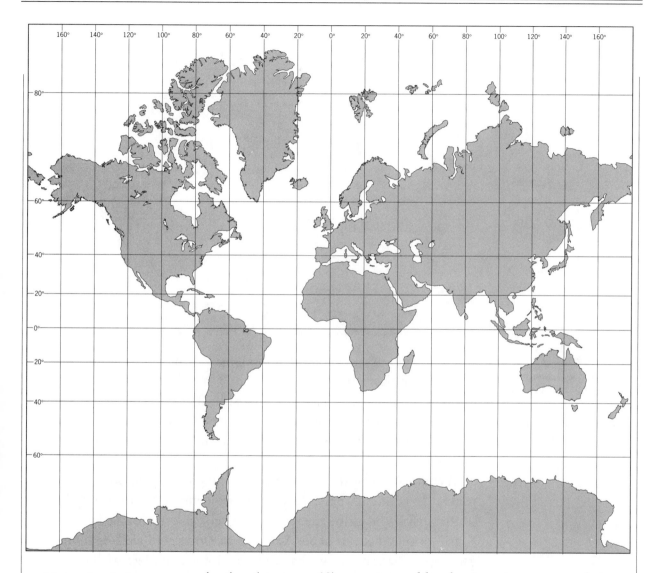

Mercator's projection greatly exaggerates the dimensions of higher-latitude landmasses, but direction is true everywhere on this map.

projection (see page 40), constructed by the American geographer Arthur Robinson, which represents the comparative sizes of countries more accurately (or rather, with less distortion). As editor of the Society's scientific journal, called *National Geographic Research*, at the time, I was invited to the news conference where the change from Mercator to Robinson was announced. When it came time for questions, a reporter from a local news organization rose and asked, "Why has it taken you so long to make this change? It seems to me that the old map reduces the size of Africa and other tropical areas,

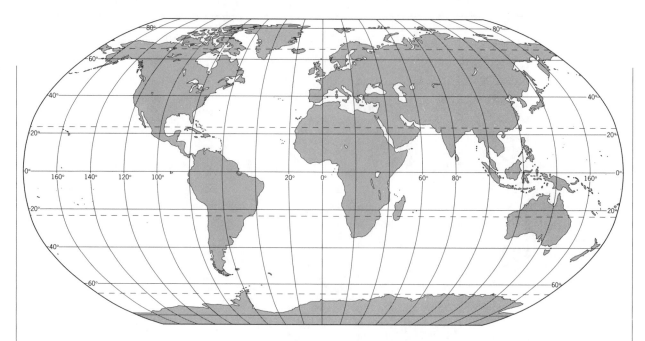

The Robinson projection substantially reduces the latitudinal size magnification. It better approximates dimension, but it lacks direction utility.

and it's a kind of cartographic imperialism!" She had a point: the Mercator projection gives not only the United States but also the former colonial countries a huge boost in size. But, as the Society's representative said in response, the biggest "loser" from Mercator to Robinson was the then Soviet Union. It "shrank" by as much as 47 percent!

What Mercator did was to project the globe's grid onto flat paper. To create his and many other projections, an imaginary light source inside the globe throws, or projects, the shadows of the global grid onto a piece of paper wrapped around it. In well-equipped geography classrooms, we have a wire globe, complete with meridians and parallels, and a sharp light we can move around inside this grid. Students then hold a cylinder of paper around the globe, or fold it like a cone-shaped hat, and watch the projections. The possibilities are endless. And so is the number of actual and potential projections in mapmaking.

MANIPULATING MAPS

By bending the grid lines certain ways, we can make areas or countries look larger than they really are, or smaller, or shaped differently. Cartographic deception—intended or accidental—is more common than you might think.

My colleague Mark Monmonier of Syracuse University has written a delightful little book about this, entitled *How to Lie with Maps*, published by the University of Chicago Press in 1991. Monmonier points out that while some cartographic distortion is actually intended to deceive and misinform, most errors on maps used by the media result from sloppy cartography, poor map design, or inexperience on the part of the mapmakers. Many of the maps we see on television, for example, are drawn by people whose background is in art, not cartography. These maps often violate all kinds of rules and conventions, because their makers never had any formal training in cartography in a Department of Geography. Every school of art or design should make a course in introductory cartography mandatory for its students. When you work with television's electronic "paintbox," such knowledge can come in handy.

And there's power in cartography. Imagine that you are opposed to the building of some installation—say a power station, or a waste dump, or a jail—in or near your neighborhood. You want to distribute a map at the meeting where the issue is to be discussed. But how to give that map maximum impact? Obviously, distance to the homes of those concerned is the key. You can display this by drawing concentric circles to represent it. And by making the circles thicker as you approach the NIMBY (Not In My Back Yard) offender, you can make the same map *much* more threatening.

Monmonier cites another kind of power wielded by some cartographers. Scanning a 1979 road map of Michigan and northern Ohio, he spotted a couple of towns with unusual names: Goblu and Beatosu. Take the road to these places, and you'll find they don't exist. The cartographer, obviously, was a Wolverine fan. When he or she was drawing that map, the Michigan-Ohio State game must have been approaching and "Go Blue! Beat Ohio State University" seemed like a good idea. Just don't plan to stop for gas at Goblu or Beatosu.

More seriously, there *is* something called "cartographic aggression." This is officially sanctioned mapmaking to proclaim possessions a country does not actually have. Before Operation Desert Storm, we became aware of Iraq's official maps, which showed Kuwait as Baghdad's "nineteenth province." Those maps, published long before Iraq's invasion of Kuwait, gave good evidence of Iraq's intentions.

Such aggression can be overt, as in the case of Iraq, or more subtle. Not long ago, I received a book from China entitled *Physical Geography of China* by Zhao Songqiao, published in 1986 in Beijing. On the frontispiece is a map of China's territory, by way of introduction. But that map, to the trained eye, looks a bit strange. Why? Because in the south, it takes from India virtually all of the Indian State of Arunchal Pradesh, plus a piece of the State of Assam. Now this book is not a *political* geography of China, nor is this matter of the Indian territory ever discussed in it. China's border is simply assumed to lie inside India, and the mountains and valleys there are discussed as though they belong to China. Make no mistake, such a map could not have been published without official approval. That map, and others like it, put the world on notice of a latent trouble spot.

THE TROUBLE WITH PLACE NAMES

I have limitless admiration for cartographers. They draw boundaries and put names on maps, go to bed at night, and wake up in the morning to find that boundaries have shifted, new countries have formed, and names have changed. Their job is never done. Prepare an atlas, and it will—especially these days—be out of date before it is published.

What is happening today is not really new. During the age of exploration, new information kept arriving on the drawing tables of European cartographers, and maps had to be revised constantly. But in the period following decolonization and before the collapse of the Soviet empire, there was a certain stability, and map changes were—comparatively—few.

Now, the rush is on again. In addition to the usual changes, cartographers face a flood of new names for cities, towns, and other geographic features in the former Soviet Union. This follows hard on the heels of the change in China from the old spelling to the new Pinyin system. And now, in addition to all those changes, whole new countries are forming: Eritrea, Slovenia, Slovakia, Macedonia.

I was working at the National Geographic Society when it adopted China's Pinyin system for its famous Atlas, and for all its other publications. In China in 1981, I was able to practice some of the new names: Beijing for Peking, Xizang for Tibet, Chang Jiang for the Yangtze River. Both in print and in the field, it took some getting used to: how about Guangzhou for Canton? While visiting the great city on the Pearl (yes, still the Pearl) River, I talked with my host about the perils of such changes. "Now, what will you Americans call Cantonese cuisine?" he asked. Good question. Guangzhouese cooking? It will never do!

At the same time, there was no hint of what was to happen in the Soviet Union, and Chicken Kiev seemed safe. But now Kiev has been renamed Kyyiv. Chicken Kyyiv?

When the National Geographic Society published the revised version of the sixth edition of its Atlas, it contained approximately 10,000 changes, yet it was out of date when it hit the streets. The flood of name changes continues. New countries and internal divisions of countries are proclaimed. There's the Republic of Ingushetia, which recently separated from the Republic of Chechnya, in Russia. There's Jharkand State, about to be proclaimed in India. Is self-proclaimed Somaliland a new country or not?

The changes continue, not only in Russia and in the other republics of the former Soviet Union, but all over the world. We seem to be in a state of transition, a time of geographic instability. In the meantime, cartographers are not likely to be out of work.

THE BOARD OF GEOGRAPHIC NAMES

Who decides when the United States, through its official publications, will accept and adopt a new name or a different spelling? This is no minor matter. Our official adoption of a name proposed by one country may profoundly upset another. American approval of the new spelling of a name may make citizens of the place in question very unhappy. Macedonia, for example, is the name of a newly independent republic—a name to which the Greeks lay claim and whose use by a neighbor enrages them. Geographic names can be sensitive issues.

Fortunately, the process of approval is done with the utmost care and consideration by an official committee whose members represent a wide range of expertise and experience. This is the United States Board of Geographic Names, an interagency committee that consists of nine members, each representing a branch of the U.S. government concerned with or affected by such issues. This nine-member Board is divided into two standing committees: the Committee on Foreign Names (four members) and the Committee on Domestic Names (five members).

The Committee on Foreign Names consists of representatives from the Defense Department, the State Department, the Central Intelligence Agency, and the Library of Congress. This group considers primarily changes in, and spelling of, country names, important internal divisions (such as the "republics" inside Russia), and international features—that is, features that extend from one country into another (such as mountain ranges and rivers) and whose names may differ on opposite sides of borders. A small army of staff researchers keeps the Committee abreast of changes made within countries, for example, from Leningrad to Petersburg, or new spellings, such as Kyyiv, which are the prerogative of the countries involved.

How does a name change or spelling change become officially approved by the United States? There are various routes. A government may send a formal communication to the U.S. Secretary of State or to the U.S. Embassy in the country, requesting affirmation of a change. If consideration of the change falls within the jurisdiction of the Committee on Foreign Names, the file is considered there and either approved or deferred, pending additional information. Eventually, the approved name is codified during a quarterly meeting of the full Board.

Many names, indeed thousands of them, are changed without such request for approval. Still, the United States needs a consistent form of these for official use. Thus it is the job of the staff researchers to comb government decrees, gazettes, maps, and other sources to secure the necessary information. This information is then collected by the Defense Mapping Agency and submitted to the full Board.

To disseminate the approved names, the Board publishes the *Foreign Names Information Bulletin*, used throughout the government and elsewhere (including atlas makers and map companies) for information on accepted spellings. Only very rarely are the names published in the *Bulletin* not immediately adopted; for cartographers, this is the ultimate source.

If you want to keep track of the world's new names and be the first on your block to spell Kyrgyzstan correctly, call 301-227-2360 and ask to be put on the *Bulletin*'s mailing list.

MENTAL MAPS

I'm sure you've noticed it: some people seem to have an innate sense of direction. They can find a street or a store with the greatest of ease. They don't miss highway exits and always know where the one-way streets are.

Others are not so lucky. They get in the wrong lane, can't remember on which side of the stadium they parked their cars, lose their way trying to find the home of a dinner host.

Geographers' research has proved that when it comes to the

maps in our minds, our *mental maps*, we are not born equal. Just as some people are color blind and others have perfect pitch, the brain's capacity to imagine our activity in a spatial way varies from person to person.

I have some evidence of this in my files. Throughout my nearly 40 years of teaching, I have asked students—not only from America, but from all over the world—to draw maps of their home city, their state, or their country (and sometimes the world) on a blank piece of paper. You would be amazed at the results. Some students can draw, from memory, a remarkably accurate map not only of their city or state, but of any part of the world. Others cannot even draw the barest outlines. Not all of this is a matter of education or exposure to geography. Indications are that our capacity for what the technical people call "spatial cognition" varies quite widely.

Perhaps you have seen those funny postcards they sell in Texas, showing a map of the United States almost completely occupied by the Lone Star State, with all the others squeezed in a narrow band against the coasts and the Canadian border. Well, I have seen this sort of thing in real life—and not as a joke. During the 1960s, when I taught at one of my favorite institutions, Michigan State University, hundreds of students came from Africa to study there. Quite a few of them took my introductory geography class. I always asked my students to draw their mental map of a continent.

Almost always, the American students drew North America, and the African students drew Africa. And almost always I was impressed by the detail African students put on their maps of Africa. But class after class, year after year, I noted something interesting: many Nigerian students drew Africa the way those Texas postcards show the United States. Nigeria would occupy almost all of West Africa and much of Central Africa, too, and the other African counties would lie squashed around Nigeria's perimeter. Nigeria, to be sure, is a large country, and in fact it is the most populous country in all of Africa. But in the minds of quite a few Nigerian students, it was also the Texas of Africa. "Well, Nigerians think big," said a student from Ibadan when I asked him about this. "Let me try drawing the map again." He did, now mindful of his earlier exaggeration. But when he was finished, Nigeria still was about twice the size it should be.

Mental maps can be improved, of course, through the study of geography. My colleague Thomas Saarinen of the University of Arizona has tested students' mental maps throughout the world, with fascinating results. When asked to draw a world map, for example, many students put Europe in the center of it, even when they don't live in Europe and weren't Europeans themselves. That is just one leftover of the educational systems spread worldwide in colonial times.

In general (this should not surprise us), we Americans are rather fuzzier, mental-map-wise, than many of our contemporaries. Does it matter? Geographers think so. As for me, I still remember that day in 1962, when, as a young Assistant Professor at MSU, I had been invited to join a group of colleagues in the State Department to discuss, with an assistant to Secretary of State for African Affairs G. Mennen Williams, a set of urgent African concerns. But the night before, President Kennedy had appeared on television with a map of Indochina, and in our Washington hotel rooms we had watched his "chalk talk" that revealed the discovery of the Ho Chi Minh Trail. The next morning at State, nobody wanted to talk about Africa. The hot issue was Indochina, and the Trail—and Laos, through which part of it lay. There was much arm waving, numerous proposals and suggestions, but the maps on the wall were of Africa, not Southeast Asia.

"May I just ask," I said as the debate swirled, "can anyone here name the six countries that border Laos? It seems to me that the layout of the region is rather important, given all these ideas."

No one could do it, and worse, no one seemed to care. "That's a waste of time," said one of my colleagues. "If we need to know that, we'll get a map and look at it."

I suggested that if you have a mistaken mental map of a place, you won't know where you're going and that if a whole cadre of decision makers had a vague mental map, we'd be in for bigger trouble than the President had intimated.

In the years since that little incident I've often thought of it and of how little we Americans, on average, really knew about the Indochina where we would wage so costly and bloody a war. Because a mental map is more than a skeleton outline. At its best, it is a store of information, not only about the layout of a place, but also about its components, its suburbs or cultures, rivers or forests. It is the spatial equivalent of our temporal, chronological knowledge, our ability to place major events in historical perspective. But our mental maps are not historic, they are current, and they guide our actions—whether it is a trip to the movies or a military campaign in a distant land.

HOW IS THE *GIS* CHANGING THE MAP?

Not everyone is familiar yet with these three letters, *GIS*—but before long almost all of us will be. That's because GIS (short for Geographic Information System) is changing the business of mapmaking and representation as never before.

For 2,000 years, maps were drawn by hand; even today, wall maps and atlas maps are prepared by cartographers bent over drafting tables, using modern versions of old-style equipment. But all of that is changing. As the versatility of computers increased, and their graphic performance capabilities grew, all kinds of possibilities opened up. Today, the map you see in your favorite magazine may well be drawn by a computer that has been instructed to manipulate information (on boundaries or resources or ethnic groups or any other spatial feature). That information comes from a Geographic Information System.

A GIS is essentially a collection of computers and programs that combine to collect, record, store, retrieve, analyze, manipulate, and display spatial information on a screen; of course the display can be printed out, so that you can still hold a paper map in your hand. Because of the huge capacity of today's computers, the amount of information in them is almost infinite. An atlas of the future will allow you to select any area of the world and bring it to the screen at whatever scale you desire. A map of the forests of Poland? The oil reserves of Venezuela? The boundary between Iraq and Kuwait? The new names of African towns? It will all be there for the asking. More on this when we look at the world of atlases and globes, but one point is clear: this is a revolutionary change. A GIS allows for a dialogue between the map and the map user; no longer is the map static and unchangeable (except through laborious alterations in subsequent editions and reprints). The map user asks for information; the computer guides the user toward answers. This is called *interactive mapping*, and it will become the cornerstone of the cartography of the future.

For example, videodisk maps will be displayed on automobile dashboards or perhaps in a corner of the windscreen, allowing the driver to see through it as he or she navigates on it. The driver can ask questions; the voice-activated computer will display the best route to the driver's destination.

All this is not to suggest that the old-style, booklike atlas will soon disappear. No computer screen can yet compete with the large, detailed spread of information on a double-page atlas sheet. But just as those atlases will retain their purpose, so will the new GIS technology transform other kinds of cartography. It's another geographic revolution, and it won't be the last.

REMOTE SENSING

Gone are the days when navigators returned to their home bases loaded with riches—and with new information to be added by cartographers to the unfolding world map. In a way, that's what has been going on ever since the Babylonians etched the first maps in wet clay about 4,500 years ago (some of these baked clay tablets survive in museums to this day, and they're worth a visit). The ancient Greeks, the Romans, the Flemish and Dutch, and British cartographers all refined the art, a refinement carried to its apogee by American cartographers during the first half of the twentieth century. But even then, most new information came from surface sources, that is, from improved ground-based equipment. Photography from airplanes helped, but only to a very limited degree.

But when the Soviets launched *Sputnik*, the space race was on, and *remote sensing*, hitherto confined to aerial photography, took off. Infrared photography had been the ultimate technique; now, new instruments were devised that could record a much wider range of the electromagnetic spectrum, that is, the electric and magnetic energy emitted at various wavelengths by all objects on the Earth's surface. Now, artificial satellites with remote-sensing systems communicate directly with computers; we don't have to wait for mapmakers to record the information. We can watch the Sahara expand and the Amazonian rainforest shrink. The satellites and computers show us things no atlas map could: changing ocean temperatures, advancing weather fronts, pollution streaks in rivers, acid-rain damage in forests.

Among the most important of these Earth-orbiting satellites is the GOES weather satellite belonging to the National Oceanic and Atmospheric Administration (NOAA). It is in a geosynchronous (fixed) orbit, so it remains stationary above the same point on the Earth's surface; from there, it observes the oceans and coasts of the United States, watching for storms. Today, television weather forecasters (some of whom were first trained as geographers) depend heavily on the data transmitted by GOES. On their televised maps, we can watch weather systems move across the country—animated cartography unheard of just a few years ago.

Also vital are four Landsat satellites that were launched between 1972 and 1982 to provide a steady stream of data about the Earth and its myriad resources. Using a battery of state-of-the-art multispectral scanners and special television cameras, Landsat's sensors have provided new insights into geologic structures, the expansion of deserts, and the growth and contraction of algae and other organisms that are crucial to food chains in the oceans. At the same time, these satellites carefully monitor world agriculture, forestry, and countless other environment-related human activities.

The world of maps is changing at breakneck speed. It has been estimated that the various nations' space satellites send back more new information to Earth in one day than was collected in a century in Columbus's time. Computers are storing all the data, but the information is useless unless it is analyzed—and unless action is taken based on the implications of the data. For all this, we need more specialists in remote sensing, in GIS technology, and in computer cartography: in short, in geography.

C H A P T E R   3

THE GEOGRAPHY BEHIND LANDSCAPE

Geographers, like most other people, like to travel, to see new places, to revisit favorite spots. For geographers, travel is also a learning experience. I'll never forget my first flight across the United States, back in 1959. While a graduate student at Northwestern University, I had taken a series of courses called The Geomorphology of America taught by a brilliant professor named William E. Powers. In those courses I not only had learned about the landscapes of this continent, but also had become acquainted with some magnificent maps, veritable works of art, drawn by E. Raisz and A. K. Lobeck. These maps showed the United States in three dimensions, not through contours or shaded relief, but by the most meticulous line work. You could see the vastness of the Great Plains and the grandeur of the Rocky Mountains, and I imagined flying across these regions by various routes. Actually, that was a good exercise. Professor Powers liked to ask examination questions such as "If you were to fly in a straight line from Atlanta to Seattle, what physiographic provinces would you cross?" And then, quite unexpectedly, I had a chance to do this in actual fact.

On a visit to the University of California at Los Angeles (UCLA) to discuss the possibility of my participating in a research project in Africa, I found myself at the window of a Constellation, flying nonstop from Chicago to Los Angeles. The weather map was promising: clear skies all the way. I had my Raisz map, a camera, and a notebook, and we departed at noon.

What followed was one of my greatest experiences. From the moment of takeoff, it was a riveting ride. I could see the Township-

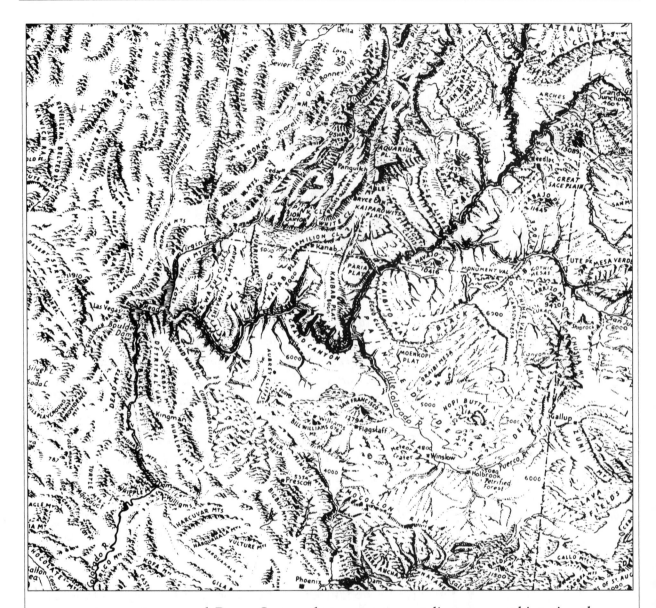

A section from one of A. K. Lobeck's remarkable maps illustrating the western United States near the Grand Canyon.

and-Range System that compartmentalizes our rural interior, the ridges and mounds left by the glaciers that had covered the Midwest, the expanses of the Great Plains, the snow-capped Rockies. I had drawn a line of flight on my map, but the captain varied our course and told us where we were. So magnificent was the view, he said, that we would do an "S" over the Grand Canyon. I saw all of Professor Powers' physiographic provinces (the term for physical regions, such as the Colorado Plateau or the Coastal Plain) and could identify and name many features we overflew. Of the several hundred pictures I took, more than a dozen have been published—and I still use several when I teach physical geography.

What a geography lesson that trip was! I saw the center-pivot irrigation circles in the dry lands, traced the Continental Divide, saw a dune landscape piled against the western slopes of the Rockies, and watched the Colorado River turn brown with mud. I thought the Grand Canyon was the most spectacular landscape I had ever seen. Ribbons of green in the dry West revealed intermittent stream drainage. Vegetation patterns changed from one moment to the next. I was only sorry when we crossed over the Sierras and descended into the vast urban area of Los Angeles. But my admiration for cartographer Raisz was never greater: those landscapes really did look like his map.

OPEN THOSE SHADES!

I've taken more than 2,000 flights since that memorable day, and many of them have given me a chance to see unusual things: the southern tip of Greenland, the volcanoes of the Aleutian Islands, the narrow entrance to the Mediterranean Sea, the ribbon of the Suez Canal, the crater of Mount Kilimanjaro, the coastline of Antarctica, and much more. I always try to get a window seat on the side away from the Sun and have been rewarded with some of the world's most magnificent sights.

But in recent years that hasn't always been easy. Now, in-flight movies require that the cabin be darkened, and flight attendants often take the liberty of lowering your window shade. Recently, on a flight from San Francisco to Miami, I compared what was happening on the screen to what could be seen outside. As we crossed Great Salt Lake and headed for the mountains, a magnificent panorama of sun-drenched snow on towering peaks and ridges unfolded. On the screen, a man armed with a knife was threatening a woman. Soon we could see the Front Range as the Rockies gave way to the Great Plains. On the screen, police cars were being demolished during another mindless car chase. As the entire Mississippi Delta came into view, etched sharply against the azure of the Gulf of Mexico, the in-flight entertainment showed a fugitive being gunned down by a blood-smeared officer; he stumbled to a ledge and then fell hundreds of feet, bounding off a rocky slope. By the time we reached Tampa Bay, it was dusk, just as the movie ended. Window shades went up; a last drink was offered.

I don't like to be subjected to images of violence, so I see far fewer movies these days than I used to. But I do love to look out the window—even the clouds or the stars can be fascinating. That's why I insist on keeping my window shade open. This world is just too wonderful to shut it out.

LAND AND WATER

About 70 percent of the Earth's surface today is water, and just 30 percent is land—and land includes the ice of Antarctica, the desert of the Sahara, the barren wastes of northern Siberia. I say today because these proportions were quite different just 25,000 years ago. We live today during a warm spell in the Earth's history, so much ice has melted and ocean levels are fairly high. But during cold periods, ocean water is taken up into ice and permanent snow, sea level drops, and long-submerged land reappears. The next time this happens, probably not very long from now, the coastline of the eastern United States will shift steadily eastward, as much as 150 miles (about 240 kilometers). The U.S. Coastal Plain will nearly double in size—and cities such as Jacksonville, Florida, Charleston, South Carolina, and Baltimore, Maryland, will lie many miles from the sea.

Any map of North America (or of any other continent), therefore, is a still picture of a changing relationship between land and water. Those shallow, now-submerged plains off the coasts are called *continental shelves*, and the continental shelf off the eastern United

DECLINE AND FALL OF SOUTHEAST TEXAS

If you believe that such matters as crustal tilting and warping are merely textbook topics, consider what is happening in Baytown and nearby areas of coastal Southeast Texas. Baytown lies near Trinity Bay, an extension of Galveston Bay on the Gulf of Mexico. With its offshore sandbars and lagoons, this would seem to be a coastline of emergence and accumulation. Instead, it is an area of subsidence, and in places very rapid subsidence. The ground is tilting downward, and the water is invading the land. The rate of decline is fast indeed—about 6 inches (15 centimeters) per year in some areas. Seawalls that once stood high above the water now barely hold back the waves. Areas that were once high and dry are now swampy and wet. Some people have already abandoned their homes and simply left, having given up the battle against the course of nature. And it is not just Baytown that is threatened. The whole Houston region of coastal Southeast Texas is subsiding, tilting southward and downward. And what the people in Baytown are facing is only the beginning.

States is one of the world's widest. The British Isles lie on just such a continental shelf, and when sea level drops again, we'll be able to drive from England to France without using the Channel Tunnel (Chunnel, as the British call it).

You can tell the extent of the world's continental shelves from many maps, because in recent times these maritime regions have become increasingly important. The continental shelves slope gently seaward until the water above them is about 600 feet (roughly 200 meters) deep. Then, the slope steepens markedly; geographers call this the *continental slope*, the real margin of the continents.

Countries with large continental shelves are often very fortunate. Oil and natural gas are found within them. Certain minerals are found on top of them. And the world's largest schools of fish concentrate above them. Not surprisingly, coastal countries want to claim as much of their continental shelves as they can.

THE OCEANS

Geography is in vogue these days as a trivia topic, and one of the most frustrating questions I hear goes like this: "How many oceans are there on Earth?"

The answers to such questions depend on the definitions of their subjects. When is a body of water large enough to be called an ocean? Are the North and South Atlantic Oceans two separate oceans, or parts of one ocean?

The trouble is, the new interest in geography is causing a lot of nongeographers to write questions that may be unanswerable as posed. Many academic geographers tend to take this sort of thing very seriously. They find fault with trivia questions, and they dislike events such as the National Geographic Society's Geography Bee, hosted by *Jeopardy* master of ceremonies Alex Trebek.

In counting the world's oceans, there are quite a few pitfalls. We all know of the three great oceans: the Pacific (largest of all), the Atlantic, and the Indian. But look southward. Do these great oceans extend all the way to the coast of Antarctica? The answer is an emphatic no. Surrounding Antarctica lies a fourth great ocean, larger than the Indian Ocean, called the Southern Ocean.

Compared to these great bodies of water, the fifth ocean, the Arctic Ocean, is little more than a large sea. Its frigid waters, clogged by ice, lie encircled by northernmost Canada, Greenland, and Russia.

So why isn't the question easy to answer?

Here's why. The reason we identify a Southern Ocean is because a great, iceberg-studded body of water, the so-called West Wind Drift, moves slowly around Antarctica from west to east. All ocean water circulates in *gyres*, giant circular currents, caused by the Earth's rotation on its axis. The southern margins of the Atlantic, Pacific, and Indian Oceans are marked by the contact between their gyres and that of the Southern Ocean. Here, the temperature of the water drops, its salinity changes, and so does its marine life. Where the Southern Ocean meets the other three great oceans, there's some mixture of water: some Pacific Ocean water gets pulled into the West Wind Drift; some Southern Ocean water gets caught up in the Atlantic Ocean's gyre.

If such a contact of gyres marks an oceanic boundary, then the same applies to the zone where the North and South Atlantic meet. There, around the equator, the differences between their waters are not so pronounced, but the two gyres *are* separate. And there is some exchange of water, just as happens in the south. Thus the North and South Atlantic Oceans are separate.

If that's the case, then the North and South Pacific are separate oceans too, because they also have different gyres. Note that in both the Atlantic *and* the Pacific Oceans, circulation in the Northern Hemisphere is clockwise, while in the Southern Hemisphere it is counterclockwise. Only the Indian Ocean, among the three equatorial oceans, is a one-gyre ocean. Its waters move in a giant counterclockwise pattern.

So if we use the criterion that separates the Southern Ocean from other oceans, then the Atlantic and the Pacific Oceans each consist of two oceans. And the trivia question's answer is seven, not five.

THE CONTINENTS

Exactly what is a continent? Here we go again! How many continents are there in the world? That depends on how a continent is defined.

The root of the word continent seems to be the Latin word for "holding together," although a predecessor of "continuous" may also have something to do with it. Here, though, is a case where human geographers and physical geographers would differ. To human geographers, Europe is a continent. To physical geographers, Europe is just part (and a small part, at that) of the Eurasian continent.

Is Australia a continent? Human geographers and physical geographers might agree here. Because Australia is so different culturally from neighboring Southeast Asia, it's a separate continent. And because Australia is also very different from its neighbors in terms of physical geography, it's a continent.

Because "continent" is so imprecise, geographers generally prefer to refer to the 30 percent of land on our planet as *landmasses*. That way, there's no doubt as to what is intended. So the Earth has six landmasses: Eurasia, Africa, Australia, North America, South America, and Antarctica.

On these landmasses lie the world's geographic realms, the cultural expression of human occupation. Europe is distinctly one of these, and so is Africa—but only Africa south of the Sahara. North Africa belongs, cultural geographically, with Southwest Asia. Today's map of human-geographic realms, as we will see in later chapters, does not match that of physical landmasses.

EARTH'S OLDEST ROCKS

The landmasses existing today look nothing like the landmasses of earlier stages in the Earth's development, but they do contain the oldest rocks known, and thus our oldest links to the Earth's distant past. Rocks believed to be somewhere near 4 billion years old have been found in three places: in Africa, Australia, and Canada.

Since the Earth is between 4.5 and 5 billion years old, there's quite a gap between the time of its formation and the evidence provided by these oldest rocks. The entire planet, shortly after its formation, was a mass of molten material of complex chemical composition. Its red-hot surface was roiled by gigantic waves of lava.

As the tens of millions of years passed, material of especially high density sank toward the center of the fiery ball, while lighter material concentrated nearer the surface. Then, when the Earth was still quite young, a huge planetoid struck it. Part of the planetoid, and part of the Earth's upper surface, congealed and would have sped off into space, but this ball of molten material was too heavy to escape the young planet's gravitational force. So it began to orbit the Earth, very close at first, more distant as time went on. That fireball, now quiet and pockmarked by asteroid hits, is the Moon.

The aftermath of that cataclysmic event has an effect on our planet to this day. The impact set the Earth rotating about five times as fast as it turns today, so fast that one day and night lasted only about five hours. Quite quickly, however, the Earth's rotation slowed, doubling the length of the day; and the Moon's orbit took it farther from the Earth's surface. All this—the planetoid impact, the Moon's separation, the rapid rotation of the Earth—set up waves of motion inside our planet, the effects of which are still felt.

Within the Earth, the heaviest material had by now (several hundred million years into its lifetime) settled in the deepest interior, a massive ball of nickel and iron whose high density and gravitational force anchored the planet. At the surface, the beginnings of a crust were forming, but time and again the light, low-density rocks that solidified were engulfed by molten rock, or *magma*, from below. Eventually, though, the cooling surface allowed patches of solidified rock to endure longer and longer, and the crust began to mature.

Imagine the scene: great volcanoes rose above the steaming surface, pouring out plumes of gases that formed a dense, heavy, carbon-dioxide-rich atmosphere. Condensation produced torrents of precipitation; much of the liquefied gas evaporated back from the still-hot surface. But eventually, liquids began to collect on lower areas of the crust, and the oceans appeared. Great earthquakes shook the surface, often opening cracks that emanated more molten rock; this boiled the ocean above. Over hundreds of millions of years the atmosphere's carbon dioxide dissolved into the cooling oceans. This created the conditions that enabled the first primitive life forms to appear. The single-celled organisms we call algae evolved at the contact point between primitive ocean and primitive atmosphere, thriving on carbon dioxide and spreading over the water. These algae, ancestors of the world's plants, began the process that would ultimately create our oxygen-rich atmosphere: they absorbed carbon dioxide and exuded oxygen.

Given this tumultuous first billion years, it is amazing that anything as old as 4 billion years is left at all. But some of the rocks that solidified as the molten surface cooled—*igneous* rocks, an ancient Greek word that aptly means "origin by fire"—have survived to this day.

In fact, the hearts of all the landmasses are made of such ancient rocks. The oldest part of North America is the vast expanse that extends across Canada from the eastern edge of the Rockies to the coast of Labrador. Geographers call such continental cores the *shields* of the landmasses, and Canada's Laurentian Shield is the core of North America.

Eurasia has several shields, as befits a landmass so large: the two biggest are the Siberian and the Scandinavian Shields. Africa is almost one enormous shield. South America has two major shields: one visible south of the Amazon River, the Brazilian Shield, and one to the north, the Venezuelan Shield. Australia is almost all shield west of the Great Dividing Range, and even Antarctica has a shield, buried beneath thousands of feet of ice.

Next time you're driving in central Canada, or if you happen to be in Scandinavia, or on holiday in East Africa, do stop at a road cut and have a look at the rocks. You'll find that these ancient rocks are made of tightly welded minerals, some of which glisten in the sun, especially the quartz. That's why igneous rocks, along with others that are formed by cooling of liquid material, are often called crystalline rocks.

Sometimes such a stop at a road cut can yield amazing results, and I often travel with a geologic hammer in my car trunk. Some rocks actually show the stresses they've undergone, and you can see how they were crushed and deformed. When you take a piece of rock and break it open to see a cleaner "face" of minerals, imagine this: that rock was formed thousands of millions of years ago, has been there all that time, and you are the first to set eyes on it. To me, it's always a thrill.

But back to the geographic question. It's becoming clear: the landmasses on which we live formed around those original shields, the oldest surviving pieces of the crust. But where, more than 4 billion years ago, did those solid patches first form? How many were there? Perhaps they formed equidistant centers of crustal cooling separated by still-boiling crust. In any case, the original continents looked nothing like those of today. The grandeur of the Earth's landscapes was yet to evolve.

THE ROCKY CRUST

The first rocks in the crust were formed by the cooling and solidifying of molten matter at the Earth's surface. These igneous rocks range in age from the very oldest to the very youngest: when lava pours from a volcano and solidifies, it, too, becomes igneous rock. That's how new land is forming on the island of Hawai'i today.

Soon after the oldest igneous rocks formed, though, they began to break down under the onslaught of rainstorms, fierce winds, and coastal waves. Grains of minerals were pried loose and these accumulated on the first beaches, in low-lying basins, and in stream valleys. Sometimes these accumulations of loose material were cemented in some way, or weighed down by overlying layers and compacted. Thus a new set of rocks developed: the *sedimentary* rocks. In road cuts and river valleys, you can recognize these quickly. They tend to be layered, or bedded, and they're often quite soft compared to the hard igneous rocks. They also have the capacity to bend, even fold like an accordion. You can see this in the Appalachians, Ouachitas, and many other places.

Other sedimentary rocks formed first as deposits on the bottom of the sea. The Great Plains was such a sea, filled with layer upon layer of sedimentary accumulation. Now these layers are rocks, but the marine fossils in them tell the story of their origins.

That's another difference between igneous and sedimentary rocks. You won't go looking for fossils in granite or basalt. But you can make some wonderful discoveries in sandstone, shale, limestone, and other kinds of sedimentary rocks.

Now to complicate matters, there's a third kind of rock in the crust. The crust is not a stable, inert layer. In fact, the crust is so thin, even after more than 4.5 billion years, that it still cracks and breaks and allows molten material to reach the surface. The crust is generally between 5 and 30 miles (8 and 48 kilometers) thick—thinner than the shell of an egg, comparatively, and much thinner than the cardboard your household globe is made of. Small wonder that earthquakes still bedevil us!

Because the crust is still active and is subject to heat from below and pressures from its own movements, rocks are continually transformed. Take, for example, a layer of sandstone, formed mainly from grains of quartz eroded from granite. Suppose a nearby eruption of lava flows over this sandstone layer, heating it up, melting many or all of its grains, and baking it into a much harder, more homogeneous rock, whose name is quartzite. In the same way, shale might be transformed into slate, limestone into marble. Such heat-baked or pressure-cooked rocks are called *metamorphic* rocks.

When it comes to *metamorphism*, the process of change in rocks, nothing is safe. Sedimentary rocks, even igneous rocks are metamorphosed. Many of those ancient granites that form the shields of the landmasses show signs of metamorphism. Even metamorphic rocks themselves (such as gneiss, schist, or quartzite) are subject to renewed metamorphism.

You can recognize metamorphic rocks because they tend to be hard, crystalline, and often shattered and lined. One effect of metamorphism is to line similar minerals up in rows during the heating process. Those lines show up in hand samples as well as broad sweeps of landscape.

THE FORMATION OF LANDSCAPE

As the crust matured, the variety of its landscapes increased. Not only do rocks differ in origin, they also react differently to the forces of erosion. When rain, streams, wind, waves, and eventually glaciers attacked the landmasses, they carved from the rocks the most fantastic sculptures. Today, we can marvel at the grandeur of the Himalayas, the drama of the Grand Canyon, the splendor of East Africa. These are the latest in an endless cycle of landscapes. The Appalachians once stood nearly as tall as the Rockies; the Andes were once a vast coastal plain, and the Amazon flowed into the Pacific, not the Atlantic.

Landscapes change not only because rocks are different, some withstanding the forces of erosion and others yielding quickly. There's another force at work. This force doesn't just carve and mold. It moves whole continents. And as the landmasses move, they crumple up into giant multiple folds like the Andes Mountains of South America, the Alps of Europe, and the Great Dividing Range of Australia.

CONTINENTAL DRIFT

When I was an undergraduate student in Johannesburg, South Africa, in the 1950s, all geography students had to take at least one year of geology. (I decided to make geology my second major, because it was so interesting to me.) There, I learned about an almost unbelievable process called *continental drift*, the movement of whole landmasses relative to each other over the surface of the planet.

My professor, a brilliant lecturer named T. W. Gevers, was one of the pioneers in continental drift research. He showed how Africa and South America had once been connected, how their coastlines, rock strata, fossils, and other properties matched across the South Atlantic Ocean. No student graduated from that department at the University of Witwatersrand with any doubt that continental drift had happened—and was still going on.

When I arrived at Northwestern University for graduate work in physical geography, I simply assumed that the conventional knowledge in Africa, and in Britain, Brazil, Argentina, India, and Australia, would already be old hat there. But I was mistaken. Northwestern's geology department held no brief for continental drift. "It's mysticism, de Blij," said its chairman, Arthur Howland, after attending a graduate seminar in which I reported what I had learned in Africa. Three years later, Howland came to my dissertation defense, that final challenge where you must answer criticism of your research before you get the

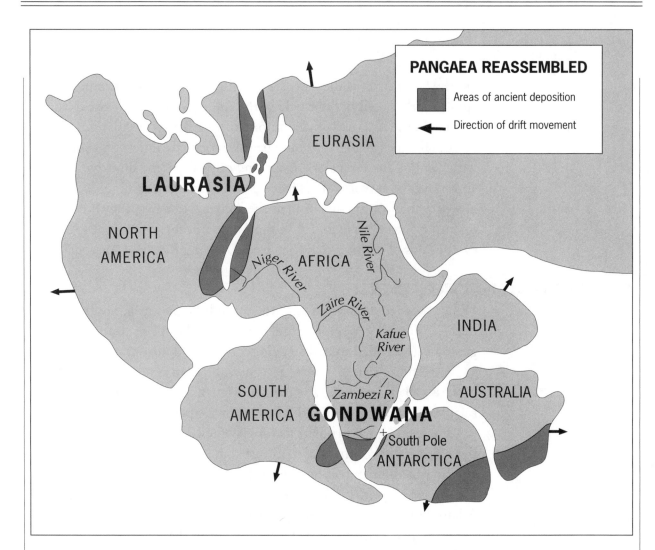

PANGAEA REASSEMBLED

Areas of ancient deposition

Direction of drift movement

EURASIA

LAURASIA

NORTH AMERICA

Niger River

AFRICA

Nile River

Zaire River

Kafue River

INDIA

SOUTH AMERICA

Zambezi R.

GONDWANA

AUSTRALIA

South Pole

ANTARCTICA

The breakup of the super-continent Pangaea began more than 100 million years ago. Note the radial movement of its remnants away from Africa and how areas of ancient deposition help us to understand where today's landmasses were once joined together.

doctoral degree. Again he attacked continental drift as a myth, and tried, unsuccessfully, to sway my committee against my conclusions, which were based in part on continental drift.

Why did it take so long for continental drift to be recognized as a possibility? Because geologists had a difficult time pursuing a research agenda set by a geographer, some scientists think; also, perhaps, because the idea was based on a spatial paradigm.

The geographer was a German climatologist, Alfred Wegener, who nearly a century ago began the painstaking job of collecting all the information he could about the opposite sides of the North and

South Atlantic Ocean—their physiography, geology, paleontology, and more. He noted landscape similarities between Canada's Nova Scotia and Newfoundland, on the one hand, and Scotland, on the other. Using the appropriate projections, he actually cut out the maps of Africa and South America and fitted them together. He correlated fossils and traced fractures in the crust from one continent to the other. He saw how the mountains at the southern tip of Africa—the Cape Ranges—extend into Argentina.

In short, Wegener amassed a huge pile of evidence, not only about the landmasses across the Atlantic, but also across the Indian and Southern Oceans. In 1914, he published a book entitled *The Origin of Continents and Oceans* and set off the debate that was still going on, at least here in the United States, in the late 1950s.

Wegener's book contained a series of maps showing how the landmasses of the world were once united in a giant supercontinent he called *Pangaea* (opposite page). About 200 million years ago, the great landmass began to crack up, producing the continents we know today. Ever since, those landmasses have been moving slowly in various directions. Africa lay at the heart of Pangaea and moved least of all. But South America drifted westward, Australia eastward, India northward, Antarctica southward. The evidence Wegener produced was overwhelming, and soon additional confirmation emerged. A South African geologist-geographer named Alex du Toit wrote a treatise on Africa and South America that all but proved Wegener's hypothesis correct. Other scientists, focusing on particular areas, added further to the weight of evidence.

HOW DID IT HAPPEN?

Geologists saw all this geographic evidence, but they would not acknowledge the possibility of continental drift until a mechanism had been discovered. And that was Wegener's problem: for all his proof of the former supercontinent's existence, he could not account for the forces that tore it apart.

Fortunately, Wegener's evidence was so strong that some geologists did pursue the notion. Some of these geologists were marine geologists. And from the bottom of the oceans came the first indication. Roughly at the center of the Atlantic Ocean (North as well as South) lies the *Mid-Atlantic Ridge*. On the map, this looks like a long, sinuous, submarine mountain chain. But the marine geologists discovered that here, molten material wells up from below the crust, solidifies, and becomes part of the ocean floor. As more material wells up, the older material is pushed aside. And this sideways push eventually drives the landmasses apart.

When Africa and South America were still united, a great crack, or fault, appeared, a giant fissure through which lava poured out. This lava hardened, but more arrived. The fault widened, and soon, geologically speaking, Africa and South America stood a few hundred yards apart. Water entered the crack, and the Atlantic Ocean was born. Eventually it looked like the Red Sea does today. But the eruptions at the fissure did not stop, and the two landmasses drifted ever farther apart. On the other side of Africa, the same thing was happening to India and Australia. Not only were new continents born from the old landmass of Pangaea, but new oceans opened up, too. Now we know that at the bottom of every ocean, there lies part of a globe-girdling system of submarine ridges. That's where the Earth's new crust is born and where the "push" of continental drift begins.

TECTONIC PLATES

From this research came another conclusion: the Earth's crust is not continuous, but broken into a number of *tectonic plates*. These are slabs of hardened crust consisting of heavier rocks at their base and lighter rocks at their top. The lighter rocks are the landmasses on which we live.

We are not sure just how many such plates there are, because more are being discovered as research goes on. There may be as few as 20 or as many as 40. For example, it was for some time believed that the African Plate consists of the African landmass plus the ocean floor all the way up to the Mid-Atlantic Ridge in the west and to the Indian Ocean's midocean ridge in the east. But now it seems that there may be an East African Plate, a West African Plate, and possibly additional plates. Several smaller plates have recently been discovered off the west coast of North America. So the map of the world's tectonic plates is still being filled in.

The crust's tectonic plates rest on a thick, molten, active layer beneath it called the *mantle*. This mantle consists of material ranging from solid (just beneath the crust) to viscous (like hot, sticky tar) and even liquid in the form of magma ready to pour out through volcanic vents. Geologists now believe that there is an upper mantle, just over 400 miles (640 kilometers) thick, that rests upon a lower mantle nearly 1,400 miles (2,250 kilometers) thick. In the upper mantle, the heat not only keeps rocks molten, but also drives them in regular cells of motion. These subcrustal cells of movement may hold the key to continental drift: if the upper mantle is in motion, that motion could drag pieces of the crust along.

SEAFLOOR SPREADING

Today, continental drift is often referred to as seafloor spreading, the name given to it by geologists. In a way, it's a misleading name. When Pangaea still existed, the supercontinent was cut by fractures called faults that split off the tectonic plates we know today. But those fractures cut across continental crust, not seafloor. Only later did the cracks open up into oceans, and the midocean ridges formed.

A better name would have been *crustal spreading*.

TECTONIC PLATE MOVEMENT

The continental landmasses, then, are carried along on tectonic plates that move relative to each other, probably dragged by heat-induced circulation cells in the mantle below. But the geologic term misleads again: this isn't just a process of spreading. The tectonic

WHEN THE BIG ONE STRIKES

The Tokyo-Yokohama-Kawasaki urban area is the biggest metropolis on earth. But Tokyo is more than a large city: it constitutes the most densely concentrated financial and industrial complex in the world. In 1993, 12 of the world's largest banks were headquartered in Tokyo. Two-thirds of Japan's businesses worth more than $50 billion U.S. are also clustered here, many with vast overseas holdings. More than half of Japan's huge industrial profits, averaging about $100 billion U.S. annually since the late 1980s, are generated in the factories of this gigantic metropolitan agglomeration.

But Tokyo has a worrisome environmental history, because three active tectonic plates are converging here. All Japanese know about the 70-year rule: over the past three and one-half centuries, the Tokyo area has been struck by major earthquakes roughly every 70 years—in 1633, 1703, 1782, 1853, and 1923. The Great Kanto earthquake of 1923 set off a firestorm that swept over the city and killed an estimated 140,000 people. Moreover, Tokyo Bay virtually emptied of water; then a *tsunami* (a seismic sea wave) roared back in, sweeping homes, factories, and all else before it. That Japan could overcome this disaster was evidence of the strength of its ongoing economic miracle.

Today, Tokyo is more than a national capital. It is a global financial and manufacturing center in which so much of the world's wealth and productive capacity are concentrated that an earthquake comparable to the one of 1923 would have a calamitous effect worldwide. Ominously, the Tokyo of the 1990s is a much more vulnerable place than the Tokyo of the 1920s. True, building regulations are stricter and civilian preparedness is better. But whole expanses of industries have been built on landfill that will liquefy; the city is honeycombed by underground gas lines that will rupture and stoke countless fires; congestion in the area's maze of narrow streets will hamper rescue operations; and many older high-rise buildings do not have the structural integrity that has lately emboldened builders to build skyscrapers of 50 stories and more. Add to this the burgeoning population of the Kanto Plain—approaching 30 million on what may well be the most dangerous 4 percent of Japan's territory—and we realize that the next big Tokyo area earthquake will not be a remote, local news story.

The Japanese would need cash to rebuild, which would force them to sell many of their worldwide holdings. This would quickly precipitate a global financial crisis. It is a measure of the interconnectedness of our world that we should all hope that nature will break its 70-year rule in the mid-1990s and prolong Tokyo's stability.

plates do spread apart in certain areas, as the African and South American Plates are doing. But elsewhere, plates are colliding. The Indian-Australian Plate, for example, is carrying India straight into the underbelly of Asia, a collision that is creating, today, the Earth's greatest mountain range, the Himalayas. The Himalayas and the Tibetan Plateau are one vast bulge created by this convergence of tectonic plates.

Tectonic plates also slide past each other. The San Andreas Fault in California marks this kind of contact between two plates: the North American to the east and the Pacific to the west. The Pacific Plate is moving northward relative to the North American Plate.

PACIFIC RING OF FIRE

The most spectacular manifestation of tectonic plate movement and contact today encircles the greatest of our oceans, the Pacific. From Chile through Central America, along the U.S. and Canadian west coast, the Aleutian and Kurile Islands, Japan, the Philippines, and Indonesia, to New Zealand and beyond lies a "ring" of active volcanoes, faults and fissures, earthquake zones, and other manifestations of crustal instability.

This is a zone of great risk, imperiling millions and killing many every year. Tokyo, one of the two largest cities on Earth, lies near the convergence of three tectonic plates and in the shadow of a great volcano. The city has repeatedly been devastated. In 1923, a major earthquake and its aftershocks killed an estimated 140,000 people. That was before a forest of skyscrapers rose above the Tokyo townscape. Over the centuries, Tokyo has been struck about every 70 years—and preparations for the next event are under way.

So many occurrences take place on the Pacific Ring of Fire that only computers can keep track of them. Landslides, mudflows, dam breaks, eruptions, gas clouds, and other death-dealing incidents mark the Pacific Ring. Still, populations keep growing and people keep coming, for while the geologic region is dangerous, the geography of it is full of opportunity. It's an old story: how quickly people forget what has happened, and how willing they are to take risks where they live.

SOME CONTINENTAL TRIVIA

Here's another geographic trivia question:

Name South America's great mountain range. Answer: the Andes. North America? The Rockies. Europe? The Alps. Asia?

THE WRATH OF PELE

Native Hawai'ians do not regard Hawai'i's volcanism as simply a geologic phenomenon. They have lived with Hawai'i's fountains of fire far longer than the white invaders have been on their islands. The goddess Pele (pronounced "PAY-lay") rules here, and she displays her pleasure or wrath through her power over the main island's volcanoes. In accordance with tradition, Hawai'ians walk barefoot on the lava to the very edges of Kilauea and Hawai'i's other steaming craters, pray to Pele, and leave fern garlands for her.

During the 1980s, plans were announced to tap geothermal energy from the interior of Kilauea. The project was expected to generate enough electricity some day to provide all the Hawai'ian Islands with power. But Pele's followers argued that such a penetration of the heart of a holy mountain would destroy the goddess herself. A confrontation developed, and it reached the courts of law. Christian missionaries long ago had suppressed Pele worship, but the geothermal project proved that Pele still has many followers. Meanwhile, the goddess seemed to prove a point by sending lava into the Royal Gardens subdivision and consuming several houses. In early 1988, the Pele Defense Fund bought a full page in *the New York Times* to state its case—and the day it appeared, Kilauea put on a volcanic display described as the most spectacular in more than a year. Native Hawai'ians were not surprised.

The Himalayas. Australia? The Great Dividing Range. Africa?

Africa, the second-largest landmass on Earth. True, there are the Atlas Mountains in the far north and the Cape Ranges in the south, but where's Africa's mountain backbone? Where are its Andes?

The answer is there's nothing comparable in Africa. And the reason has to do with Africa's position in the middle of Pangaea. Note that the Earth's great mountain ranges are situated *away* from Africa: the Andes in western South America, but the Great Dividing Range in eastern Australia. While the landmasses were carried away from Africa and their leading edges crumpled up into great folded mountain chains, Africa rotated slightly but moved very little laterally. If you take a wax pencil and outline the world's major mountains on a globe, they form a circle around Africa that only touches Africa in the north. The Atlas, actually, are part of the chain of which the Alps and the Caucasus are also parts.

This was one of Wegener's earliest arguments, and he had it right.

WHAT WEGENER DIDN'T KNOW

Wegener believed that the Earth's landmasses had been one vast continent until, about 200 million years ago, Pangaea split apart as continental drift began.

Today, we know that the current phase of seafloor spreading is only the latest in a sequence that may have begun as soon as the first primitive landmasses formed on the crust. There is evidence that the landmasses have converged into one giant continent on as many as three or four occasions *prior* to Pangaea, only to come apart under tectonic forces. So the present process may be the fifth phase of continental drift in the Earth's history. Some geologists speculate that, in another 40 or 50 million years or so, the landmasses will stop moving outward into the Pacific, will remain stationary for a long period, and then begin to converge again, presumably on the heart of the Land Hemisphere, Africa. By the time they approach Africa, we would not be able to recognize them. An Andes-like mountain range may form along the east coast of North America. In the west, the mountains of today may be eroded into a flat plain. All landscapes are temporary.

IS AFRICA SPLITTING UP?

Right now, though, the landmasses are still spreading outward, away from Africa. That leads to a question: will Africa itself split up?

A look at a map suggests just such a scenario. The midocean ridge that marks the Indian Ocean margin of the African Plate runs right through the Gulf of Aden and seems to push into Ethiopia. There, the midocean ridge becomes a two-walled valley, a kind of trough with steep sides that gave Ethiopia its old name—*Abyss*inia.

Follow that trench southward, and you can see that it leads through Lake Turkana, then splits in two. The western trough contains several of East Africa's great lakes, including Lake Tanganyika. Then the valleys meet again and fill with the waters of Lake Malawi. It almost seems as though the piece of East Africa from Ethiopia to Swaziland lies poised to separate from the rest of Africa. Is this what it looked like when Africa and South America began to separate?

There's no conclusive answer to this question yet, but those steep-sided valleys that suggest this to be the case have another great significance. These often lake-filled troughs are called *rift valleys*, and only Africa has them in such magnificent dimensions. Here in East Africa, perhaps in these very rift valleys and on their step-like slopes, humanity evolved. Here, in the shadow of Mount Kilimanjaro, human genes were formulated. Here stood our earliest ancestors, gazing at landscapes that have not changed greatly since hominids arose: great, snow-capped volcanoes rising over vast, forested plains broken by steep-sided valleys filled with lakes. Over several million years of human evolution, climates changed: forests gave way to grasslands, coolness replaced tropical warmth, and drought withered the lakes. Some of our ancestors could not adapt, and their lineage ended; others adjusted, overcame, and thrived—to enjoy the return of milder conditions.

I'll never forget the first time I set eyes on this scene, more than 30 years ago, because it changed my life. In one moment, at dawn on the north side of Mount Kilimanjaro, I realized that this, somehow, was home—a second home, a place to which I would always want to return. Since then, I've been back dozens of times, occasionally with students. I've watched their reactions, and some of them were affected the same way I was. Several go back whenever they can, just to be there, not knowing exactly why they have to go—but go they must. I've been all over the world, but no place has ever touched me the way East Africa has. Maybe it's because we all started there. We are, you know, all Africans by origin.

YOU'RE MOVING!

There's no longer any doubt that we all live on continental rafts carried on tectonic plates moved by forces in the mantle below. In earthquake zones, these movements are betrayed by earthquakes and tremors, but over far larger areas, we do not notice continental drift at all.

CAUSING A RIFT

My doctoral dissertation was to be based on fieldwork in Swaziland, a country in southeastern Africa wedged between Mocambique and South Africa, and supervised by a famous geomorphologist, L.C. King of the University of Natal in Durban.

Upon arriving in Durban, I visited Professor King and reported that my preliminary research suggested there might be a rift valley in eastern Swaziland. "Then the first task you have is to prove that this valley is no old river valley," he said. "You'd better dig some soil trenches and get profiles of the ground."

Several weeks later, I was in the Swaziland Lowveld, accompanied by a dozen Swazi helpers. We walked through the densely overgrown area, hoping to find a place where we could dig without having to cope with the roots of acacia trees and bush. One morning we found it: a wide, open stretch of grassland with nary a tree on it. I laid out a plan of action, and soon we were digging our trench. Rhythmic song filled the air and, as the hot day wore on, beer flowed. By late afternoon, we had dug a considerable ditch, evinced by a long pile of soil across the open patch.

Around 5 P.M., as we were sitting on our mound, we heard the sound of an engine. It grew louder. Presently, a small plane appeared overhead. That was unusual; this was remote country. We waved enthusiastically. The plane disappeared.

A few minutes later, it came back—just over the treetops. Then it skimmed the grass, and if we hadn't dived into our furrow, it would have taken our heads off with its wheels. As we watched the pilot turn for what looked like another run at us, my eye caught a rag hanging from a tree in a corner of the grassy patch. Only then did I realize what had happened; that rag served as a wind sock, and our patch was a landing strip. We had dug up the local airfield!

So much, I thought, for geographic awareness. My first question should have been Why *are* there no trees on this large, open field?

And yet, we're constantly on the move. We now have equipment to measure what Wegener surmised, which is the motion of the landmasses in certain predictable directions. Even while the North American Plate and the Pacific Plate are sliding past each other, the whole combination is moving slowly into the Pacific. If you are reading this in Chicago, or New York, or Dallas, you're moving— about 1 to 1.5 inches (2.5 to 4 centimeters) per year in a generally westerly direction. Ten years from now the walls of the room you're in will be a foot from where they are today.

That doesn't seem like much, but convert it to centuries, millennia, and millions of years. A million years isn't much in the lifetime of the Earth, but in that million years North America moves a mil-

WEATHERING AND EROSION

Exposed rocks of the Earth's crust are under continuous attack by numerous processes, many of which can be grouped under two categories: weathering and erosion. Weathering is the disintegration of rocks by mechanical or chemical means but without significant movement of the fragments produced. Erosion involves both the breakdown of rocks (by water, ice, or wind) *and* distant removal of fragments created by weathering and erosional processes.

An example: alternate heating by the Sun and cooling at night can weaken a rock at its surface and break loose some of its mineral grains. This is weathering; the particles may roll from the rock surface to the ground nearby. Later, a rainstorm causes a rush of water that carries the grains into a stream and then into a river. This is erosion.

lion inches or so, or more than 15 miles (24 kilometers).

Has the rate of movement been the same ever since Pangaea split up? There is reason to believe that movement was much faster early in the breakup of the supercontinent and that it slowed down over time. This is one reason geologists believe that the process may eventually stop, to be followed by a turnaround that will see the landmasses move back together. In technical terms, continental drift may be cyclic.

UNDERSTANDING LANDSCAPES

To begin to understand how landscapes are fashioned from a variety of rocks by a variety of forces, we must know how the crust formed and reformed, how rocks were created and then transformed by heat and pressure, how continents moved and collided. Now we know why the Himalayas, the highest mountain range in the world, are still growing, while the Rockies, still spectacular, no longer mark a zone of plate collision and are now being worn down. We know why America's active volcanoes are found in the west, where plate contact prevails, rather than in the more stable east, which is more or less in the middle of the plate. In geographic terms, we have a better understanding of why things are where they are.

But there's still more to it. Not only are the Earth's landscapes fashioned by tectonic forces, they're also sculpted by weathering and erosion. In deserts, where daytime heat is fierce and nighttime cold can be frigid, rocks are exposed day after day to great temperature extremes, and gradually they weaken, their grains loosened by continuous expansion and contraction. Those desert dunes are piles of loosened grains, shaped into characteristic forms by the action of wind. Along the coasts, the waves exploit softer rocks to carve bays and inlets, while longshore currents try to close them off with sandbars. In the mountains, rushing streams erode deep valleys that often mark the weaknesses of the rocks.

Only when we look at the whole picture, the forces from below as well as above, and the action at the surface, is it possible to look out of your car or airplane window and say "Here's where we are on the map, and here's how this landscape formed." It can make travel an entirely new experience.

CHAPTER 4

DOING SOMETHING ABOUT THE WEATHER

Geography is about places. And places are characterized by their weather. If a change of job were to force you to move to a new city, one of your first concerns is likely to be What's the weather like there?

Geographers, therefore, are ardent students of weather and its causes. Many of us get our first glimpse of physical geography when we enroll in a course that deals with this topic. Of course, the weather is only one aspect of a place's total environment. Plants, soils, local crops, and other factors also come into play. But among these, the weather is dominant. That's why weather forecasters are fixtures on our television news programs.

Quite a few television meteorologists, as they are often called, had their first experience with weather study in a geography class. A couple of my own students have gone on to television stardom, making a career out of weather commentary. They really know their regional geography, rattling off states and cities and foreign countries as they tell the weather story. Some other weather forecasters had little or no background in this area when they were asked to pinch-hit in front of the weather map. For them, a good geography book is the quickest way to get up to speed.

WEATHER FORECASTING

Even with today's sophisticated knowledge of the atmosphere and our high-technology equipment to monitor its every murmur, weather forecasting remains as much an art as a science.

The American public today is probably more aware of weather conditions than ever before—witness the stampede to supermarkets and hardware stores when a snowstorm or hurricane is forecast. This is due in no small part to the improvement in the news media's weather coverage in recent years. On television, the inarticulate little bow tie man, with his broken car antenna pointing at a messy, postage stamp-sized map, is long gone. Today we are far more likely to see weather reporters with meteorology degrees who use animated maps, color radar, and frequent references to jet stream behavior; and with millions of U.S. homes now able to receive The Weather Channel via cable television, viewers are better informed than ever. The daily press was slower to follow, but when *USA Today* introduced its full-color weather page in the mid-1980s, every major newspaper in the nation swiftly upgraded its weather reporting and graphic presentation.

The World Meteorological Organization, an agency of the United Nations, supervises the World Weather Watch, a global network of more than 10,000 weather stations and moving vehicles. Nearly 150 nations participate in this monitoring system, which coordinates and distributes weather information from its processing centers in Washington, D.C., Moscow, and Melbourne, Australia. Every three hours, beginning at midnight Greenwich mean time (GMT), a set of standard observations on the local state of the atmosphere is taken around the globe and reported to those centers from about 4,000 land-based stations, more than 1,000 upper-air observation stations, and at least 5,000 ships, aircraft, and satellites in transit. High-speed computers then record, manipulate, and assemble these data to produce *synoptic weather charts*, which map meteorological conditions at that moment across wide geographic areas.

Although reliable weather instruments have been available since about 1700, comprehensive, simultaneous observation could not be attempted before the invention of long-distance communications. The arrival of the telegraph in the 1840s provided the initial breakthrough, but it wasn't until a half century later that the

U.S. Congress, responding to a number of weather-related disasters (such as the great blizzard of 1888), funded the establishment of a monitoring network by the Army's Signal Corps. By the turn of the twentieth century, these efforts had intensified with the founding of the Weather Bureau in Washington, D.C.

After the nation entered the Second World War in 1941, weather forecasters were required to expand technologies and hone their prediction skills for the military services, and these were applied to civilian forecasting after 1945. As postwar technology advanced ever more rapidly, including the use of orbiting satellites after 1958, the Weather Bureau kept pace. In 1965, a major reorganization of the environmental agencies of the federal government resulted in the creation of the National Oceanic and Atmospheric Administration (NOAA), and the Bureau was incorporated as the National Weather Service.

WEATHER AND CLIMATE

There's a difference between weather and climate. Weather refers to a set of conditions prevailing at a given place at a particular time. When, for example, television weather personality Spencer Christian of *Good Morning America* reports that it's a comfortable 75 degrees with a relative humidity of 50 percent and a slight breeze from the south, he's making a *weather* statement.

When we take all the weather information we have about a place and combine it, we get an overall picture of year-round conditions there. That total record is the climate of the place—the average of all the available data. So when Christian refers to Chicago's very warm summers and windy, cold winters, he's talking about the *climate* there.

That's why our geography course is called Climatology and not Weatherology. We're interested in the overall workings and patterns of climate, although, of course, we also need to learn why and how weather conditions develop as they do.

GEOGRAPHY OF CLIMATE

What can geographers tell us about climates? In fact, many geographers work in highly specialized, technical aspects of spatial meteorology, where they deal with complex mathematics and computers. But this does not mean that the field of climatology has been bypassed. Here are some questions that keep us busy:

• *How do weather and climate affect individuals, communities, and societies over the short and long term?* Many years ago, geographers were tempted to speculate that certain climates are superior to others. Ellsworth Huntington was one of the leaders in this work, and he wrote some highly speculative and tendentious books such as *Civilization and Climate* (1915) and *The Pulse of Asia* (1907). Critics argued that Huntington's work could contribute to notions of superior and inferior societies (and races), and this branch of human geography fell into disrepute. But the fact is that the questions Huntington raised have not been satisfactorily answered, and they are being researched again—more objectively this time.

• *How have global and regional climate changed over the past several thousand years?* Using all kinds of records, such as tree-ring data, winegrowers' logs (which go back centuries), glacial movements, diaries, ice cores, and other sources, climatologists try to reconstruct our climatic past. They've concluded, for example, that the Earth was exceptionally warm toward the end of the first millennium A.D., during what they call the Medieval Optimum.

• *What do maps of world climates look like?* Can we draw maps that tell us in an instant that the climate of, say, Santiago, Chile, is similar to that of Florence, Italy? Such maps can reveal in an instant what migrants and travellers can expect of their new environments, what farmers can grow, or what natural vegetation will predominate.

• *How strong is the evidence supporting predictions of a possibly catastrophic greenhouse warming during the coming half century?* While the 1980s were comparatively warm, they may not have signaled the greenhouse onset claimed by some scientists. Many geographers working on this topic are not prepared to endorse the greenhouse warming alarm.

OUR SPEEDING PLANET

The Earth orbits the sun at an average distance of about 93 million miles (150 million kilometers), on a near-circular path that carries us 585 million miles (940 million kilometers) in approximately 365 days. This translates to nearly 67,000 miles (108,000 kilometers) per hour, a dizzying speed by earthly standards. We don't notice it, of course, because all else—our atmosphere included—moves along with us.

Astronomers studying the orbits and paths of asteroids, meteors, and comets warn that we are not safe from collisions. Some have even suggested that we should apply star wars research to develop the capacity to intercept and pulverize any objects that might collide with the streaking Earth—just as the original plan was to intercept Soviet missiles.

THE ROTATING EARTH

Not only does the Earth speed through space, it also rotates rapidly. One rotation is completed in 24 hours. Consider this: since the circumference of the Earth is roughly 25,000 miles (40,000 kilometers), a person at the equator moves at a speed of well over 1,000 miles (1,600 kilometers) per hour. In the middle latitudes, this unnoticed speed is down to around 500 miles (800 kilometers) per hour, and if you stood straddling the North or South Pole, you'd merely turn around during the 24 hours.

CORIOLIS FORCE

We may not notice the effect of the Earth's rotation as we go about our daily business, but it affects us nonetheless, and in some very consequential ways.

The Earth's rapid rotation affects all moving objects on the Earth's surface and above it. It affects you when you drive your car, planes when they fly on a given course, space shuttles when they take off, winds when they blow, ocean currents as they drift, rivers where they flow.

The scientist who first analyzed the effect of the Earth's rotation on all moving things was a Frenchman, Gustave-Gaspard de Coriolis (1792–1843). Coriolis found that then-existing laws of motion required the insertion of an inertia force into the formula. Furthermore, Coriolis realized that this force would operate in opposite ways in the Northern and Southern Hemispheres. Ever since, it has been known as the *Coriolis Force*.

The practical effect of what Coriolis discovered is to create opposite effects in the Northern and Southern Hemispheres. In the Northern Hemisphere, all moving objects are pushed to the right of their intended direction. In the Southern Hemisphere, moving things are pushed to the left. Here's one very obvious consequence: in the North Atlantic Ocean, water circulation moves in a clockwise direction, westward from Africa toward the Caribbean, northward in the Gulf Stream, southward along the northwest African coast. But in the South Atlantic Ocean, water moves counterclockwise: southward in the Brazil Current, eastward in the West Wind Drift, and northward along the African coast in the Benguela Current.

So it is in the other oceans and in the atmosphere. Air circulation, too, is affected by the Coriolis force. In the Northern Hemisphere, low-pressure systems (the kind that include fronts and storms) circulate counterclockwise. In the Southern Hemisphere, these circulate clockwise. An American weather forecaster who took a job with an Australian television station would have a tough time adjusting to this!

DID CORIOLIS DOOM KL 007?

On September 1, 1983, a South Korean commercial airliner flying from Anchorage, Alaska, to Seoul strayed from its assigned course, crossed into Soviet territory, and was shot down by a Soviet jet fighter, with a loss of 269 lives.

Why KL 007 should have strayed off course has remained a mystery. It was not, as the Soviets alleged, a spy plane. But if its computer program, inserted in Anchorage, failed to adjust for the Coriolis effect, the plane would have drifted to the right of its intended course, to just about the degree where its fatal encounter took place.

IS CORIOLIS IN OUR GENES?

We may not notice the effect of the rotation of the Earth in our daily routines, but don't underestimate what this force is capable of. For example, if we were anatomically perfectly symmetrical and we were blindfolded and asked to walk a straight line, we would (on a perfectly flat surface) walk in a large circle, clockwise in the Northern Hemisphere, where we'd be pushed to the right, and counterclockwise in the Southern Hemisphere.

Could Coriolis have something to do with our overwhelming right-handedness? The great majority of people live in the Northern Hemisphere, where right-handedness is dominant. As humans, we have occupied lands in the Southern Hemisphere only recently. Could our righthandedness have something to do with our geographic distribution and, over millions of years of human evolution, our exposure to rightward pressure?

MYTH OF THE DRAIN

There's a popular tale that suggests the Coriolis Force causes bathwater to go down the drain counterclockwise in the Northern Hemisphere and clockwise in the Southern. Other factors, however, overpower Coriolis in bathtubs. You'd have to have a tub the size of a football field, with a drain in the center and a stopper that doesn't guide the water when it's pulled, to make this work. In your bathroom, the slope of the bath, the way you pull the stopper, and the inevitable motion in the water prior to release all make a myth of this "evidence."

STILL...

We may be able to see the Coriolis Force at work when large crowds of people walk long distances. In Britain, where the rule is to keep left, a minority habit the British exported worldwide to their colonies, there are signposts, even lines on sidewalks, urging people to walk to the left. But people seem to move to the right anyway. I've spent quite a bit of time in Edinburgh, Scotland, on assignments. Every time I've been there, overlooking Princes Street, I've noticed that people there seem to walk on the right as they promenade by the thousands along the wide sidewalks. On the other hand, in Southern Hemisphere Buenos Aires, Argentina, the rule is to keep right, but try that on fashionable Florida Avenue. There, they seem determined to keep left.

EARTH AND SUN

What stirs the atmosphere and the oceans into action, of course, is our Sun—and the Earth's tilt relative to it. The Earth's axis of rotation is presently inclined 23.5 degrees from the vertical. That is, without this tilt, the Sun's hottest rays would always fall on the equator, and there would be no swing of the seasons. But because of it, the Northern Hemisphere is bathed in the Sun's maximum warmth during the six months from April to September, and the Southern Hemisphere gets its turn from October to March.

We have all seen the effect of this as the Sun's angle rises from March 21 onward, to reach its highest position in the sky on June 21. By September 21, it's down to where it was in March, and dropping, until December 21, when people enjoy the warmth of summer in the Southern Hemisphere, while we plunge into winter, with the Sun's angle at its lowest and the Sun's warmth at its weakest.

The Sun's maximum heat rays, therefore, sweep back and forth across the Earth each year (that is, each revolution around the Sun). This sets up all kinds of movements in the air and in the water.

Facing Page: *An idealized version of the global atmospheric circulation pattern showing the major pressure belts, the cell-like airflows that develop between them, and the Coriolis deflection of surface winds.*

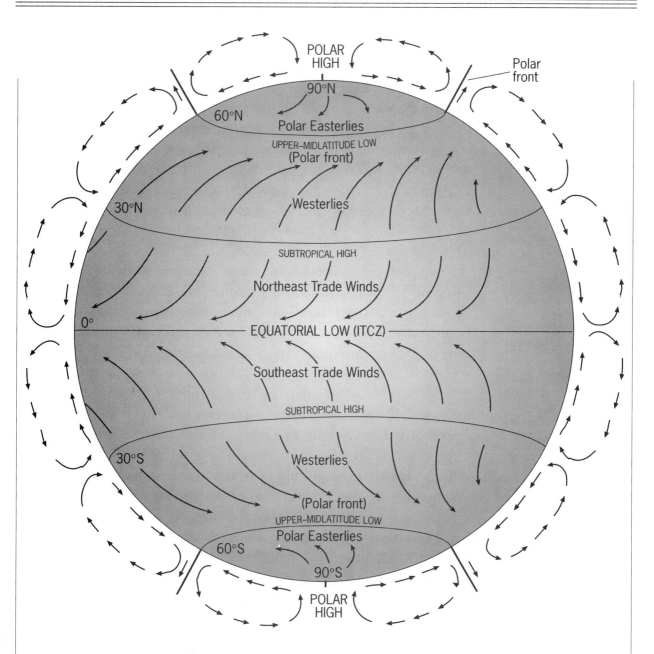

POLAR HIGH

Polar front

90°N

60°N

Polar Easterlies

UPPER–MIDLATITUDE LOW
(Polar front)

30°N

Westerlies

SUBTROPICAL HIGH

Northeast Trade Winds

0°

EQUATORIAL LOW (ITCZ)

Southeast Trade Winds

SUBTROPICAL HIGH

30°S

Westerlies

(Polar front)

UPPER–MIDLATITUDE LOW

Polar Easterlies

60°S

90°S

POLAR HIGH

These movements, in turn, are influenced by the Coriolis factor. As a result, we can make maps of oceanic and atmospheric circulations—small-scale maps, of course, to which there are many exceptions. But the overall patterns strongly influence our daily and seasonal lives.

THE EARTH IN PROFILE

The combination of differential hemispheric heating and planetary rotation propels a global system of atmospheric circulation consisting of persistent wind and pressure belts. Air has weight, and when it's warmed from below, when the Sun heats up a patch of surface, it tends to rise. The rising air vacates its place near the surface, causing nearby air to take its place. When that air flows in, the Coriolis force bends its path, to the right in the Northern Hemisphere and to the left in the Southern.

The equatorial zones are the Earth's warmest regions of fast-rising air, billowing clouds, condensation, and copious rain. Because air tends to rise there, there's not much steady direction of wind flow in equatorial areas. On the water, this creates an especially difficult zone of passage for sailing vessels, hence the term for these areas—the doldrums.

Air is drawn in toward the equatorial zones from the north and the south. But as it flows along the surface from north to south, the air is diverted by Coriolis, and assumes a northeast-to-southwest direction. And air from the south, flowing northward, is bent west-ward, ultimately moving from southeast to northwest. These are the all-important trade winds that allowed the sailing ships of old, once they'd managed to cross the doldrums, to reach their destinations.

When heated air rises, it cools, its moisture content condenses, and rains fall—hence the high precipitation of equatorial areas and their luxuriant plant and animal life. But the reverse is also true. Once up toward the top of the *troposphere*, as the life layer of the atmosphere is called, the cooled, dried-out air begins to sink toward the surface again. That happens in subtropical latitudes, and since this sinking air is dry, it yields no rain. That's why the forests of the equatorial zone such as those of Zaire yield so quickly to the desert, in this case the Sahara.

This is where we in the middle latitudes come in. The subtropi-cal air, once it hits the surface, flows outward in two directions: equatorward, to become part of the trade winds, and poleward. In the Northern Hemisphere, air flowing toward the North Pole, deflected by the Coriolis Force, will bend toward the right (that is, the east). That makes these winds the *Westerlies*, since a wind is always named for the direction *from which* it comes.

That's why, when the Cubs are rained out in Chicago on a given afternoon, you may want to think twice before buying tickets for that night's Detroit Tigers game, to the east of Chicago. But if an afternoon game is rained out in Cleveland, the night game at Chicago's Comiskey Park isn't threatened. Here in the middle latitudes, weather tends to move from west to east—generally. It's all part of the great Earth-Sun system.

WEATHER SYSTEMS

True, most of us live in the Westerly wind belt, but this doesn't mean that the wind in Kansas City or Chicago or Pittsburgh always comes from the west. What it means is that weather generally moves, in our latitudes, from west to east.

In the Northern Hemisphere trade winds, weather moves, usually, from east to west. This explains the tracks of major hurricanes: they start near the West African coast, cross the Atlantic in the trades, become affected by Coriolis and curve to the right (north), and, if they survive long enough, get caught up in the Westerlies and blow into the mid-Atlantic.

A hurricane is a weather system, and an especially intense one. By international agreement, hurricanes are called typhoons west of 180 degrees in the Pacific Ocean and cyclones in the Indian Ocean. The Earth's entire atmosphere is churned by weather systems, most of them, fortunately, less dangerous.

On the map, weather systems are represented by various symbols. Just as contour lines represent highs and lows on the surface, lines marking points of equal atmospheric pressure—*isobars*—delineate weather systems. And just as with contour lines, you can see at a

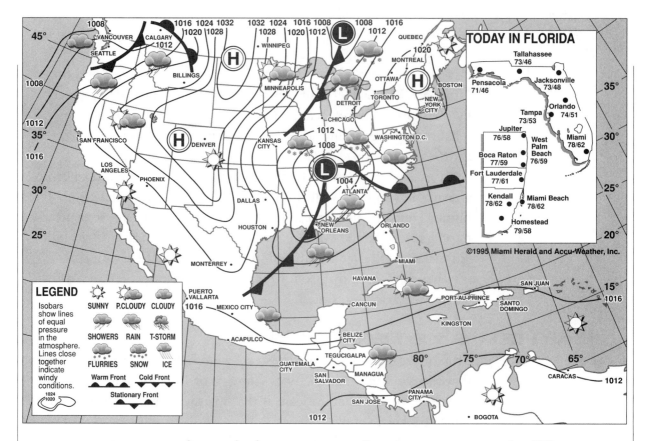

A February 1995 weather map in the Miami Herald *courtesy of Accu-Weather, Inc., State College, Pennsylvania.*

glance whether pressure gradients are steep or gentle. When contour lines are bunched closely together, you're looking at a scarp; when isobars are close together, you may be looking at a hurricane.

When air rises, it's especially light, compared to nearby air. So you'll find lots of low-pressure systems in the tropics, where air tends to rise over the sun-heated surface (a hurricane begins as a low-pressure system).

When air becomes cold and dry, it tends to become heavy, for example, over the interior of Canada during a winter day. This creates a high-pressure system. When the Canadian Express hits us in midwinter, you can be sure that it's a high-pressure system. Clear skies and hard, frigid winds are the norm.

If your newspaper or your favorite television weatherperson have a good daily weather map (above), it will show "lows" and "highs" competing for space. These highs and lows tend to be approximately round, with the isobars forming concentric circles. But note that the air circulation differs: in highs, the air moves clockwise; in lows, it moves counterclockwise (in this hemisphere).

It's a bit unfortunate that *all* low-pressure cells are called cyclones, like those severe tropical storms in the Indian Ocean. This can be confusing, although Indian Ocean cyclones are comparatively rare. Ordinary low-pressure cells, or cyclones, abound. High-pressure cells, are called anticyclones, which leaves no room for doubt.

Midlatitude low-pressure systems often become battlegrounds between different kinds of air. Because they are lows, air flows into them, not out, as in the case of highs. And that means that air from the Gulf of Mexico flows in from the south, even as air from Canada flows in from the north. That sets up zones of contact between these contrasting masses of air. Those zones, often marked by clouds, rain, thunder, and even tornadoes, are called weather fronts. Midlatitude cyclones, therefore, are complex weather systems marked by colliding air masses.

Where the cold air advances against retreating warm air, the front is a cold front, and cold fronts produce the most severe weather—high winds, thunderstorms, and hail are not unusual. These can be especially violent during the spring and fall, when the contrasts between northerly and southerly air masses are especially strong. On the weather map, cold fronts are marked by tooth-like symbols, sharp points pointing in the direction of the front's movement.

A warm front tends to be more benign. The moist, warmer air gradually gains on the retreating cooler air, mixing with it, condensing the moisture, and producing steady rain and, often, foggy conditions. The map symbols are appropriately rounded.

Consider the problems a weather forecaster has. When a low-pressure system advances on an area, its path may be generally eastward, but it may move northeast, east, or turn southeast. Its fronts can weaken or strengthen in very short order. Thunderstorms or even tornadoes can break out when a few hours earlier there was no sign of them. And low-pressure systems and their fronts have the disagreeable habit of slowing down or even stopping altogether, ruining the promise of a sky-clearing passage to make way for a high—and a sunny weekend. Monday, my students tell me, is a favorite day off for weather forecasters.

IN THE UPPER AIR

The weather we experience at the surface is greatly influenced by what happens in upper layers of the troposphere, miles above the ground. Just as there are weather systems at the surface, there are systems in the upper troposphere, and these can determine what happens to us at the bottom of this ocean of air.

Here are just two examples. You will have noticed that many a tropical storm develops and matures but does not turn into a hurricane. Usually this means that the incipient hurricane has been beheaded by a countercirculation in the upper air. For example, if a low-pressure cell (which is how a hurricane starts) drifts under a high in the upper troposphere, the reverse circulation in that upper-air high will counter the surface low and weaken it.

On the other hand, if a tropical storm, with its well-established low center, drifts under a low in the upper air, the two matching circulations strengthen the system—and a hurricane may be in the making.

Another upper-air phenomenon that affects us at the surface is the jet stream. Actually, there's more than one jet stream, but the one that affects us most is the midlatitude jet, a tubular snake of air that zigzags eastward at an altitude between 20,000 and 40,000 feet (6,000 and 12,000 meters). The jet stream develops at upper-air contacts between contrasting air masses, and its position often defines the transition (at the surface) between warm, south-

TORNADOES

No weather phenomenon can match tornadoes for destructive power. A narrow column of rapidly spinning air, a tornado can destroy all in its path as it moves across the countryside. Tornadoes appear along cold fronts; they develop in association with severe thunderstorms; and they emerge in the fringes of hurricanes. They strike on land as well as at sea (where they are usually milder and named *waterspouts*.) The interiors of large landmasses are the most vulnerable to their destruction.

A tornado's dimensions are modest: the whole spinning tube of air is between 330 and 1,650 feet (100 and 500 meters) in diameter. Inside the tube, barometric pressure may be significantly below that of the surrounding air. Meanwhile, the walls of the tube are spinning at speeds ranging from about 100 to 450 miles (150 to 700 kilometers) per hour. As a result, a tornado destroys in two ways: even if its high winds do not knock down a structure, the low pressure at its center can cause a building to explode.

A tornado begins its life as a rapidly rotating funnel of air at the base of a thundercloud. Eventually this funnel reaches the ground, where it may either sweep along or "hop" from place to place along a relatively straight path (which is why some houses are left untouched while others are demolished during the passage of some tornadoes). Their awesome, often dark color is caused by the vegetation, loose soil, and other objects sucked into the center tube of this gigantic vacuum cleaner.

In the United States, tornadoes frequently develop from squall-line thunderstorms over the Great Plains, particularly in the north-south corridor called Tornado Alley that extends through central Texas, Oklahoma, Kansas, and eastern Nebraska.

ern weather and cold, northern conditions. For instance, if your weather forecaster's map shows the jet stream in, say, December to extend from San Francisco to Denver, Chicago, and then to Atlanta, and out to sea, and you live in the Northeast, get ready: winter's here. You're on the north side of the divide between southern warmth and northern cold.

But the jet stream is capricious. The forecast may call for several days of continued cold, yet the weather warms up. More likely than not, the jet stream has taken a turn northward. What we experience on the ground is often determined way up in the air.

OCEAN AND LAND

The globe-girdling wind and pressure belts that move hurricanes from east to west and cold fronts from west to east do more than drive the weather. They ensure the interaction between land and

HURRICANES, TYPHOONS, CYCLONES

Every year, in the tropical oceans between the approximate latitudes of 5 degrees and 20 degrees north and south, tropical depressions form that develop into tropical storms and, under certain favorable conditions, into hurricanes (as they are called in the Atlantic Ocean and in the Pacific east of 180 degrees), typhoons (their name in the western Pacific), and tropical cyclones (in the Indian Ocean, especially in the Bay of Bengal).

These storms all begin as low-pressure circulation systems. When such a pressure cell develops into a hurricane, it is almost perfectly circular in shape. It has a steep pressure gradient that leads to a central (core) low, where pressure sometimes drops below 950 millibars when standard sea level pressure is 1,013.2 millibars. This generates an intense air circulation around the hurricane's core, wifth surface winds reaching 125 miles (200 kilometers) per hour and even higher. The system grows to a diameter averaging 300 miles (500 kilometers), but hurricanes more than twice this large have been recorded. Wind strength increases toward the center, but the core of the hurricane, its eye, is calm. The eye measures 3 to 9 miles (5 to 15 kilometers) across; here, the air descends from high altitude, is warmed adiabatically, and rejoins the circulation. Temperatures in the eye can be as much as 18 degrees Fahrenheit (10 degrees centigrade) higher than in the other segments of the hurricane.

Hurricanes tend to move slowly and often erratically, although they drift westward and northwestward in the North Atlantic Ocean's trades and then curve northward and even northeastward as they penetrate the westerly wind belt. This path takes them through the Caribbean Sea; occasionally, they curve into the North Atlantic and threaten the U.S. East Coast, but at other times they enter the Gulf of Mexico and strike the coast of Mexico or the Gulf states.

The wind damage done by hurricanes is aggravated in coastal areas by the water swept up by the storm. Often the storm surge does more damage than the wind itself, although hurricanes are normally attended by other dangers as well, including tornadoes set off by the incredible turbulence in the system.

ocean that is crucial to our survival on this planet.

The North and South Atlantic Oceans show how. Both oceans have gyres that move water from equatorial latitudes toward the poles and from the poles back toward the equator. Obviously, water flowing from equatorial latitudes poleward is warm, and water flowing from polar latitudes equatorward is cold or cool. Warm water tends to flow past east coasts, and cool water flows past west coasts.

Superimpose our global wind belts, and we see that the trade winds have the effect of carrying the warmth and moisture off east

coasts over the adjacent lands, producing the high rainfalls of
Caribbean Middle America and northern South America. But where
the Westerlies hit southwestern Africa and northwestern Africa, they
first cross the cool waters of the Benguela and Canaries Currents,
lose their moisture, and arrive onshore pretty dry. That's why there
are deserts on the coast in Namibia, Morocco, northern Chile, and
Baja California.

Of course there are always exceptions. Western Europe is mild
and moist, for its latitude, because the North Atlantic Drift carries
the Gulf Stream's warmth across the North Atlantic, warm
upwelling water reaches the surface offshore, and the Westerlies
do the rest. Our Pacific Northwest and coastal British Columbia
are moist because high mountains intercept the air off the cool
ocean and wrest lots of moisture from it. But the global pattern of
precipitation, overall, is explained by the combination of ocean circulation and atmospheric wind and pressure belts. And behind all
this, let us not forget, is the force of Coriolis.

A MAP OF WORLD CLIMATES

All the weather factors combined—the sweep of the Sun's energy,
the rotation of the Earth, the circulation of the oceans, the movement of weather systems in alternate zones, the rush of air in jet
streams—produce a pattern of climates that may, on the map,
look complicated at first but is actually remarkably simple. This
is one of those maps that's worth a million words, because it
allows us, at a glance, to determine what climate and weather are
like anywhere on Earth.

We owe this remarkable map (page 97) to the work of
Wladimir Köppen (1846–1940), who devised a scheme to classify
the world's climates based on indices of temperature and precipitation. Since his time, there have been many efforts to improve on
Köppen's classification and regionalization of climates, but nearly a
century after its creation, and despite the greater accuracy of climatic data today, it has stood the test of time.

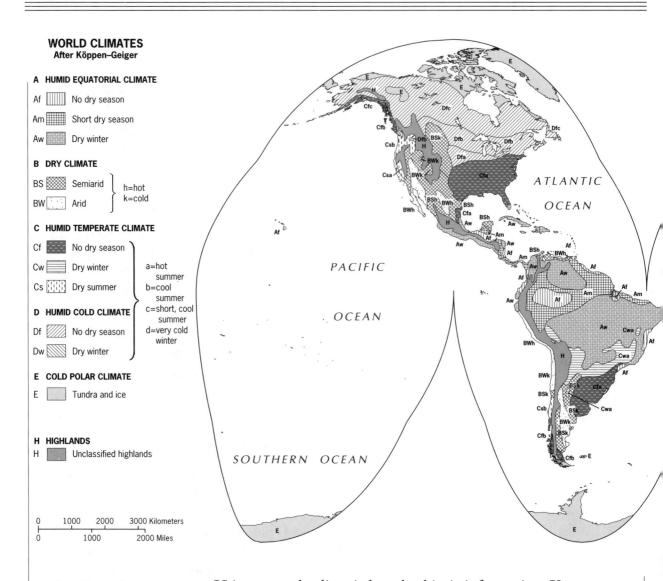

WORLD CLIMATES
After Köppen–Geiger

A HUMID EQUATORIAL CLIMATE

Af   No dry season

Am   Short dry season

Aw   Dry winter

B DRY CLIMATE

BS   Semiarid   } h=hot

BW   Arid   } k=cold

C HUMID TEMPERATE CLIMATE

Cf   No dry season

Cw   Dry winter   a=hot summer

Cs   Dry summer   b=cool summer

D HUMID COLD CLIMATE   c=short, cool summer

Df   No dry season   d=very cold winter

Dw   Dry winter

E COLD POLAR CLIMATE

E   Tundra and ice

H HIGHLANDS

H   Unclassified highlands

0   1000   2000   3000 Kilometers

0   1000   2000 Miles

A world map showing the regional classification system first developed by Wladimir Köppen.

Using not only climatic but also biotic information, Köppen established six major types of climate. He used letters to identify these on his map:

A: Equatorial, Tropical, Moist

B: Desert, Dry

C: Midlatitude, Mild

D: Continental, Harsh

E: Polar, Frigid

H: Highland

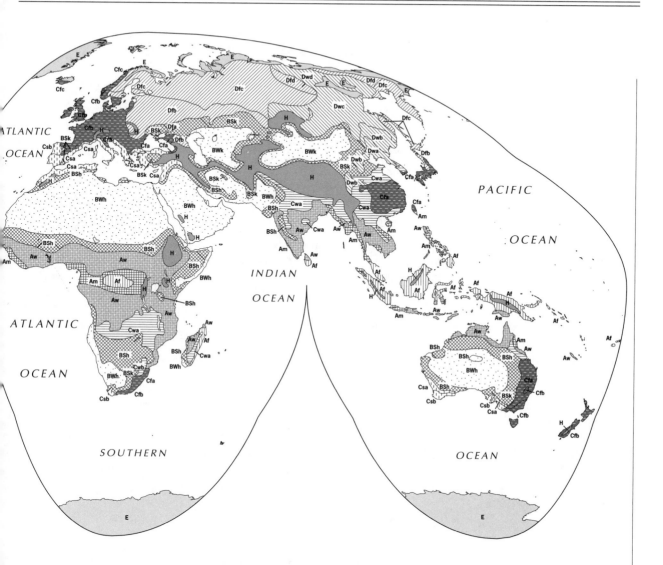

Köppen added other letters to provide more detail within each of these climatic regions, but even without these, his map is very useful. Take, for example, the climate prevailing over most of the southeastern United States, Cf on the map. If you're familiar with that climate and its local weather (for example, in Atlanta, or Nashville, or Charlotte, North Carolina), you'll have a pretty good idea of what it's like in much of eastern China, in southeastern Australia, and in a large part of southeastern South America. Chicago's climate is a Df climate, with colder winters and hot summers; you'd experience

WINEGROWING AND CLIMATE

One of the most interesting forms of agriculture is the growing of grapes (viticulture) for the purpose of making wine (viniculture). This is an ancient pursuit, vineyards having stood in Mesopotamia and pharaoic Egypt. Later, winegrowing diffused to Crete and Greece, into the Roman Empire, and (through the word of Roman winegrowers) into France, Germany, and even England. After Columbus's journeys to America, the first vines were planted in Mexico, from whence the industry spread into California. The colonists carried viticulture to southern South America, to South Africa's Cape, to Australia, and to New Zealand.

When grapes are cultivated for winemaking purposes, a complicated process unfolds. It is not just a matter of bringing in a large harvest of ripe fruit, as in the case of other crops. A large harvest, in fact, may mean inferior grapes and poor wine. Viticulturists must learn to prune their vines to keep the harvest limited. And the timing of the harvest is crucial. Too early: the grapes may be too acid and the wine sharp and unpleasant. Too late: there may be too much sugar and a flabby, unappealing wine.

The vine is a hardy plant, but the grape is a fragile fruit. Summer warmth and sunshine—and very limited rainfall—are needed; winters must be cool but not too severe. Slopes must be turned to the sun just right; high winds, hail, and other hazards can doom a vintage. Humidity, causing rot, is a constant threat in many vineyards. Viticulture and physical geography are closely associated.

Italy, base of the Roman Empire, remains the world's largest producer of wine. France, where the Romans introduced viniculture, is the second-largest producer, and French premium wines continue to command the highest prices. Spain ranks third, although the former Soviet Union may produce as much wine as Spain in an average year, or even more. Argentina, largest producer in the Southern Hemisphere, follows.

Certain wines are closely identified with their regional sources: Champagnes (France), Sherries (Spain), Ports (Portugal), Chiantis (Tuscany, Italy) among them. A wine can be a telling representative of the culture of a place, its climate, soil, traditions, and preferences. It all starts in the vineyard, in the hands of a specialized farmer.

something close to it in Moscow and Sapporo (northern Japan). The map shows you instantly where the "real" rainforest climate prevails, where the world's deserts are, and where the best grape-growing areas lie. Note that the Cs climate, a mild climate with a dry summer (which is denoted by the small letter s), occurs not only around the Mediterranean Sea (and thus in the famed wine countries of France, Italy, and Spain), but also in California, Chile, South Africa, and Australia. So you know what to expect in Rome, San Francisco, Santiago, Cape Town, and Adelaide.

Of course it's possible to take parts of this world map and draw it at a much larger scale. That takes away our ability to make world-wide comparisons, but it increases the amount of information that can be provided for a smaller region. Travel kits almost never include such a detailed climate map. They should.

CLIMATE AND SOCIETY

Ever since Köppen published his map of world climates, geographers and others have been intrigued by the apparent spatial relationship between certain climates and certain successful, powerful societies. Equatorial and tropical climates, apparently, do not favor the countries over which they prevail; none of the world's major powers, present or recent, lie here. Desert climates also do not appear to be conducive to big-time status. The same seems true of high-latitude, polar climates.

Are the mild midlatitude and the harsh continental climes the best our planet has to offer? Huntington had no doubt about it. "The people of the cyclonic regions," he wrote in 1940, meaning the midlatitude cyclonic zones, "rank so far above those of other parts of the world that they are the natural leaders...[they lead in terms of productivity, but] their greatest products are ideas and the institutions to which these give rise. The fundamental gift of the cyclonic regions is mental activity...."

Such observations made Huntington famous; they also got him into a lot of trouble. By extension, we might conclude that racial groups living under cyclonic conditions would be superior to others; and from this, it was but a short step to Nazi master-race philosophies. *Environmental determinism*, the idea that environment (chiefly climate) controls the capacities and destinies of societies, quickly got a bad name.

That was unfortunate, because Huntington had raised one of geography's central questions, and in the context of his time, he made enormous contributions to our understanding of climatic change and its impact on human societies. These days it is fashionable to join the chorus of disparagement, but many who do so have never read Huntington's last major (and monumental) work, *Mainsprings of Civilization*, which appeared in 1945, two years before his death. *Time* magazine, in its review of this book, proclaimed it comparable to one of the greatest of historical works, Arnold Toynbee's 12-volume *Study of History*.

Huntington argued that the seasonality of midlatitude climates had favored peoples who remained for many generations under these demanding, yet stimulating regimes. He explained the rise and fall of societies in terms of the "sweep" of climate change. Köppen's map, he frequently said in his lectures, should be seen as a still photo of a changing Earth—it represents the way things are today, but not as they were yesterday, nor as they will be tomorrow. When the Maya civilization rose to greatness, when West African states arose and prospered, when the Middle East and North Africa soared to intellectual eminence, it was because climate induced it.

Such arguments were bolstered by research results from psychology and physiology. Huntington saw proof of his notions in school examination results, in medical records, in factory productivity studies, and in mortality figures. "The well-known contrast between the energetic people of the most progressive parts of the temperate world and the inert inhabitants of the tropics," he wrote, "is largely due to climate."

Small wonder that such environmental determinism suffered the fate it did. But the fact is that many questions arising from Huntington's work remain unanswered to this day. We now know much more than was known in Huntington's time about the changeable nature of our natural environment; Huntington described climatic change as cyclic, and on this point he was right. Over the past thousand years, global climate has swung back and forth, cooling, warming up, cooling off again, and, since the mid-nineteenth century, warming up once more. It was the discovery of the current slow warming trend that helped give rise to the greenhouse warming hype that has dominated the headlines for nearly a decade now. Associated with these temperature fluctuations were changes in regional precipitation regimes: some places got drier, while others got moister. We can link some political events and economic circumstances to these climatic swings, but the evidence is weak yet and the time span is still short. What we need is a series of Köppen maps beginning about 10,000 years ago. Armed with these, we would be able to understand human history a great deal better than we do now.

THE GREENHOUSE WARMING MANIA

We geographers have been teaching and writing about the greenhouse effect for more than three-quarters of a century. We haven't always gotten the message across. When I graded my first examinations at Miami, a student wrote that the greenhouse effect was named after its inventor, Walter Greenhouse, who was a biologist interested in mosses.

The Earth's atmosphere lets the Sun's shortwave radiation through, resulting in the warming of the rocks and water at the surface. In turn, the heated surface sends out longwave radiation, but this energy cannot pass through the air without warming it. So the lower atmosphere is heated from below, much as the glass of a greenhouse allows light in but won't let heat out.

Given the revolutions the Earth makes around the Sun, its rapid rotation, and the mobility of its atmosphere, it's not surprising that this greenhouse effect varies over time. There are times when the Earth gets so cold that ice sheets spread deep into the middle latitudes; at other times, the Earth is so warm that its plant and animal life breed prolifically. When the dinosaurs were enjoying their apogee, the Earth was warm, sea level stood high, swamps proliferated—until, one unpleasant day, one or more asteroids plunged into what is now the Gulf of Mexico near Yucatan, and put an end to all this.

Today, we're enjoying unusual warmth too, but in a different context. Unlike the dinosaurs, which faced nothing of the sort, we humans have emerged during an *ice age*. The Earth has undergone five or perhaps six of these cold spells during its lifetime. The previous one happened before Pangaea split apart, more than 200 million years ago. Whatever the cause or causes may be, an ice age makes the Earth cool and warm up again in rapid succession. During our ice age, often called (not totally accurately) the Pleistocene Ice Age, advances of the ice, called *glaciations*, have followed each other about every 100,000 years. During the last glaciation, ice stood as far south as the Ohio River, and that was just 12,000 years ago. Much of Canada still lay under the ice as recently as 9,000 years ago.

The emergence of modern human civilization—states, cities, crops, livestock, religion, the works—has taken place during a warm spell between glaciations, about the 30th such warm spell during the current glaciation, which has now lasted about 10,000 years.

Our current warm spell has witnessed the explosion of human population from a few million when the ice retreated to nearly 6 billion today. In fact, some scientists are suggesting that we humans, by our numbers and our prodigious capacity to pour pollutants into the atmosphere, are warming the air even more than nature had in mind. As a result, it is argued, we are risking a catastrophic temperature rise during the twenty-first century, causing melting of the remaining ice on Antarctica and Greenland, elevated sea levels, and submerged coastal plains and coastal cities.

When I was a young faculty member, the scientific literature was full of dire warnings, too—of the opposite kind. All kinds of evidence was indicating that the current warm spell would soon end (these last from 9,000 to 13,000 years on average, and we're beyond 10,000 years right now). I remember one film I used, made by a consortium of British universities. As the narrator described the situation, make-believe snow began to fall from the studio ceiling. The calendar said that it was July; the place, London. It had all the drama of a high school production.

Still, there was a serious side to this. *Science* magazine carried scholarly articles predicting soon-to-come glaciation. Research funds were poured into the global-cooling arena by the millions of dollars. The popular press warned of advancing glaciers and restricted living space. Some of the world's great grain-producing areas would dry up under desiccating Arctic winds.

And now, less than 30 years later, the field has been reversed. Now greenhouse warming is the rage, and a great number of research projects I see refer to the contribution this or that work will make to the impending greenhouse-generated crisis. Now it's not advancing glaciers, but rising sea level that drives the research agenda.

What to believe? You will recall that some scientists proclaimed the 1980s the warmest decade on record. Before long, New York would be the Venice of the Atlantic, and we'd all be paddling boats from the fourth floor of the Hilton to the fourth floor of the Plaza. You'd take the Fifth Avenue ferry to Wall Street.

Fortunately, the evidence is weak. True, the 1980s were warm in North America and in parts of Europe, but check the records in the Southern Hemisphere. There, they were having some unusual, not to say unique, cold spells. There was a major, unprecedented snowfall in Johannesburg and surrounding areas of the Transvaal in South Africa. It was much colder than usual in southern South America. I was in Sydney, Australia, in January 1992 and one day watched the mercury plunge into the 50s—the coldest ever for that date. In short, while some scientists were promoting notions of anthropo*genic* warming of the atmosphere, some others, notably geographers, were noting a certain anthropo*centric* quality.

Other aspects of the evidence also troubled geographers. Although huge quantities of atmospheric data are now generated by satellite, surface records still provide key information. And many recording stations simply are not reliable over the long term. Take, for example, weather stations that once stood far from the built-up city where, as your television weatherperson always tells you, temperatures are higher. Urban sprawl has now overtaken many of these stations. They cannot be moved, or the record will be interrupted.

But the influence of the urban surroundings creates higher readings, suggesting regional increases that may not actually occur.

What to make of this? Let's take the long-range geographic view. Various kinds of evidence do suggest that there has been a very slight rise in global temperature since about the 1850s. But this is not the first time over the past 1,200 years or so (or even the second) that such a trend has taken place. In her marvelous book, *The Little Ice Age*, published by Macmillan in 1989, Jean Grove marshals a vast array of evidence to indicate that the Earth, over the past millennium, has warmed and cooled repeatedly. In the period before A.D. 1000, she reports, it was warm over much of the globe. Seas were comparatively calm, and settlements took hold on some long-inhospitable shores, including those of Iceland and Greenland. Grains grew on fields of both northern and southern Iceland. Sea level was high; the Dutch were having a tough time of it, furiously building dikes to control the frequent floods. Grapevines thrived in Britain, and wine was shipped even to France!

Viking explorer Leif Eriksson and his party used the calm waters of the North Atlantic to make it to North America. At the other end of the Earth at virtually the identical moment, Polynesian canoeists we now call Maori reached—for the first time—New Zealand. That was hardly likely to be a coincidence. What, other than environment, would have delayed the arrival of such accomplished seafarers in so large and attractive a land? About 1,000 years ago, the Earth was probably warmer, and the seas calmer, than today.

What happened next? We should pay attention, because the end of this Medieval Optimum, as geographers call that salubrious period, may mirror the end of our own current warm spell, which we might call the Industrial Optimum, given the timing of its occurrence.

No, the warming trend did not continue. Rather, it came to an end in a period of climatic extremes that included unprecedented storms, cold waves in some areas, warm spells in others, huge rainfalls in some places, and devastating droughts in normally moist

THE DROUGHT OF '88

The drought that plagued so much of the United States in 1988, the worst since the mid-1920s, was a classic example of this major environmental problem. At its peak in late summer, extreme and severe drought conditions covered more than 40 percent of the nation. For the B climate areas, which occupy most of the conterminous United States west of 100 degrees west longitude, these dry spells have always been part of the gamble of living and farming there. Even in the Mediterranean climate of the southwestern coastal zone, rainless summers are a constant reminder that water resources must be managed wisely because recurrent shortages have been part of California's modern settlement history. But what made the 1988 drought one of this country's most memorable environmental events was its grip on the normally humid Cfa climate zone that blankets the southern two-thirds of the country's eastern half.

Unlike desertification, which is a process of human degradation that affects many B climate environments, drought is a natural hazard that recurs in seemingly irregular cycles. Specifically, a drought involves the below-average availability of water in a given area over a period of at least several months. The conditions that constitute a drought are clear enough: a decrease in precipitation accompanied by warmer than normal temperatures and the shrinkage of surface and soil water supplies. If the drought reaches an extreme state of development, all of these conditions intensify further and may result in spreading grass fires or forest fires, as well as the blowing away of significant quantities of topsoil by hot, dry winds.

However dramatic it may become at its height, a drought has no clear beginning or end—it develops slowly until it becomes recognized as a crisis, and it tends to fade away as more normal moisture patterns return. Because droughts encompass no spectacular meteorological phenomena, they were not the subject of serious scientific study before the 1970s. But the Sahel disaster of two decades ago in northern Africa finally aroused the interest of climatologists, whose subsequent research is beginning to provide a clearer understanding of droughts and their often far-reaching consequences.

areas. The settlement on Greenland was extinguished. That of Iceland died out soon thereafter. The vine disappeared from Britain. Food shortages afflicted Europe. Disorder and migration were rife. These, indeed, were Dark Ages.

You may have noticed it in the press and on television: the growing number of "unusual" environmental events of the 1980s and 1990s. A hundred-year storm on the U.S. East Coast in 1992. A so-called blizzard of the century in March 1993. Major hurricanes in Florida and Hawai'i after decades of quiescence. In 1994, torrential rains in Egypt's Eastern Desert damage tombs not affected by moisture in 3,000 years. Look at the longer term: the disastrous drought along the southern fringe of the Sahara in the 1970s (which made Sahel a household word and added "desertification" to our daily lexicon). From frequent and severe El Niño Southern Oscillations (unusual heat distributions and water flows in the Pacific Ocean off South America) to withering droughts in interior Asia, the environmental seesaw that marked the end of the Medieval Optimum may be upon us again.

That's the real worry: that while we pursue the agenda of global warming, unprecedented environmental instability may lie ahead. None of this means that we should not heed scientists' warnings to reduce our pollution of the atmosphere and inhibit the destruction of the ozone layer; although variations in ozone may have natural as well as anthropogenic causes, also related to the volatility of the global environmental system in transition. It does mean, though that we should contemplate more than a world of enhanced greenhouse warming. We should prepare ourselves for a time when the challenge will come from more than this single direction.

When someone asks me about the importance of geography in our educational makeup, I cannot think of a more important dimension than this. Greater geographic awareness would have lessened the chance that our attention could have been so completely diverted toward a single, unproven issue. To quote my colleague Patrick J. Michaels in the National Geographic Society's scholarly journal, *National Geographic Research*: "The true legacy of [the] global warming [notion], which the data argue is much more benign than it is perceived to be, will be the destruction of the public's faith in science. Tragically, it will be noted by historians in the 21st century, that even by the mid-1980s, the data had indicated that the then-popular vision of climate catastrophe was a failure."

C H A P T E R   5

CITIES

Cities are where the action is. Most are dirty, dangerous, expensive, polluted, crowded, rest-less, noisy, nerve-racking. They're also, depending on where you look, beautiful, impressive, historic, energetic, productive, cultured, and awe-inspiring. I'll never forget my first look at New York, which is all these things and more. All my life (which then amounted to a mere 20 years), as long as I had known of New York, I had wanted to get there, to walk the avenues and see the famous skyscrapers. I had arrived in Baltimore by ship the day before, after 30 days at sea from what was then Portuguese East Africa. Next morning, I was on the train to New York, seated strategically by the window on the right-hand side. My eyes were glued to the horizon, but in New Jersey the conductor briefly diverted my attention. Then I looked again—and there it was, the Empire State Building, the Chrysler, and the others. I probably shouldn't admit this, but I actually rose and took off my cap. If people hadn't been looking at me, I probably would have saluted. As it was, I had tears in my eyes.

Cities attract people like magnets. Every day of the year, tens of thousands of people move to the world's burgeoning cities. Many are larger, in terms of population, than whole countries. There are more people in Tokyo than in Denmark, Norway, Sweden, and Finland combined. There are more people in Mexico City than in all the countries of Central America (Guatemala, Belize, El Salvador, Honduras, Nicaragua, Costa Rica, and

Panama). There are more people in São Paulo than in Venezuela.

Consider this: in the year 1800, less than 3 percent of the world's population lived in towns or cities larger than 5,000. In 1900, the figure was still only 14 percent. But today, more than half the world's population resides in what we call *urban* settlements. By the year 2000, if present trends continue, it will be 55 percent. From 3 to over 50 percent in less than two centuries!

FUTURE MEGACITIES

By the year 2025, United Nations studies suggest, Mexico City will have been joined by at least a dozen other cities with populations over 20 million, including São Paulo, Cairo, Lagos, Karachi, Bombay, Calcutta, Delhi, Dhaka (Bangladesh), Shanghai, and Djakarta (Indonesia). All of these cities lie in the less developed world, in the countries with less urbanization; many lie in countries much poorer than Mexico. The greatest challenges posed by the urban spiral lie ahead.

People migrate to cities on the basis of pull factors that are more imaginary than real; their expectations of a better life mostly fail to materialize. Particularly in the less developed realms, but in the industrial cities of developed regions as well, the new arrivals (and many long-term residents, too) are crowded together in overpopulated apartment buildings, dismal tenements, and teeming slums. Such arrivals come from other cities and towns and from the countryside, often as large families; they add to the cities' already substantial rate of natural population growth. No housing expansion can keep up with this massive inflow. Huge new shantytowns develop almost overnight, mostly without the barest amenities. However, in-migration is not deterred, and multiplying millions of people will spend their entire lives in urban housing of wretched quality.

Miserable living conditions for urban immigrants notwithstanding, the world's cities continue to beckon. Cities are humanity's crucibles of interaction, opportunity, achievement, innovation, and progress. They have always been the nuclei of power, foci of culture, and centers for the arts and sciences. Philosophers, composers, architects, writers, scientists, artists, and religious leaders choose cities as their home base, contributing to the excitement and ferment that great cities generate. The world's greatest orchestras, museums, monuments, engineering feats, and other outstanding human accomplishments are concentrated in large metropolises. Many of the best universities are located in and around major cities. Skyscrapers, subways, elevators, domed stadiums, and countless other innovations have emerged in the city. Nowhere is the variety of available goods and services greater than in the metropolitan centers. Life in these crucibles of humanity may be problematic for many, but that does not deter the next wave of migrants. The city is where the action is and where the opportunities are, real or perceived. The spiral of urbanization, set in motion thousands of years ago, is far from spent.

WHAT IS A CITY?

There's no doubt: Tokyo, Mexico City, and São Paulo are cities. They're huge agglomerations of tens of millions of people packed tightly into a small area, living in structures ranging from high-rises to shantytowns, and sustaining themselves in an urban economy. But smaller urban (that is, nonrural) places are cities, too. How many thousands of inhabitants does it take for a place to be a city? There's no worldwide agreement on this. When countries take their censuses, they do not all use the same criterion for urban living. Some take it to mean a place with 5,000 residents or more. Others put it at 30,000, as in Japan, or 10,000 in Spain, 2,500 in the United States, and as few as 200 in Denmark. That, unfortunately, makes urban comparisons difficult.

POPULATIONS OF THE WORLD'S
LARGEST CONURBATIONS
FROM 1990 TO 2000
(in millions)

Rank 1990	Rank 2000	Urban Area	1990	1995	2000
1	2	Tokyo-Yokohama-Kawasaki, Japan	27.1	27.9	28.7
2	1	Mexico City, Mexico	20.9	24.5	29.6
3	3	São Paulo, Brazil	18.1	21.7	26.1
4	4	Seoul, South Korea	16.7	19.4	22.4
5	8	New York-New Jersey, U.S.A.	14.6	14.7	14.7
6	9	Osaka-Kobe-Kyoto, Japan	13.8	14.1	14.5
7	7	Shanghai, China	13.0	14.0	15.2
8	5	Calcutta, India	11.7	13.1	15.9
9	6	Bombay, India	11.7	13.0	15.3
10	12	Buenos Aires, Argentina	11.5	12.2	12.9
11	10	Rio de Janeiro, Brazil	11.4	12.8	14.3
12	13	Moscow, Russia	10.4	10.7	11.1
13	14	Los Angeles Area, U.S.A.	10.0	10.4	10.7
14	11	Cairo, Egypt	10.0	11.2	13.2

Reports and estimates of urban area populations vary quite widely. Other sources may cite different figures.

Here's another problem. When we say that Tokyo is the world's largest city, we really mean the Tokyo *urban area*, or to use another geographic term, the Tokyo *conurbation*. Over the past century, Tokyo has grown so large that it has merged with its growing neighbors Yokohama and Kawasaki. So when we speak of Tokyo, we really mean Tokyo-Kawasaki-Yokohama—plus a few other towns now engulfed by Tokyo's urban sprawl. Mexico City, on the other hand, is one large city. It has its neighborhoods, and it has grown to incorporate some small settlements around it, but it is one huge city. Now, if we compare *just* Tokyo to Mexico City, then Mexico City is the world's largest city. But if we compare total conurbations, then Tokyo is still the largest though Mexico City, at current growth rates, will overtake Tokyo before long.

Geographers employ some commonly used terms to distinguish,

MEXICO CITY: WORLD'S LARGEST BY 2000

The Mexico City conurbation has borne the brunt of the recent migratory surge toward urban areas. With a total population of 22.4 million (second in the world in 1994), it is already home to nearly one out of every four Mexicans, and it continues to grow at an astonishing rate. Each day, about 1,000 people move to Mexico City; when added to the 1,000 or so babies born there daily, that produces a staggering addition of approximately 750,000 people every year. However, birth rates in urban Mexico are higher than the national level; with half its population 18 years of age or younger, some demographers have forecast an astounding total of between 40 and 50 million residents for greater Mexico City by 2010. In any case, by 2000 Mexico City will have become the world's largest single population agglomeration, surpassing metropolitan Tokyo.

Even in a well-endowed natural environment, such an enormous cluster of humanity would severely strain local resources. But Mexico City is hardly located in a favorable habitat; in fact, it lies squarely within one of the most hazardous surroundings of any city on earth—and human abuses of the immediate environment are constantly aggravating the potential for disaster. The conurbation may be located in the heart of the scenic Valley of Mexico, but serious geologic problems loom: the vulnerability of the basin to major volcanic and seismic activity (such as the devastating earthquake of 1985), and the overall instability resulting from the weak, dry lake bed surface that underlies much of the metropolis. Water availability presents another problem in this semiarid climate. Dwindling local supplies must be augmented by drinking water transported from across the mountains, an enormously expensive undertaking that requires the accompanying growth of a parallel network to pipe sewage out of the waste-choked basin.

Mexico City's appalling air pollution, however, poses the greatest health hazard and is conceded to be the world's most serious. It is exacerbated by the thin air, which contains 30 percent less oxygen than at sea level (the city's elevation is 7,350 feet [2,240 meters]).

For the affluent and for tourists, Mexico City is undoubtedly one of the hemisphere's most spectacular primate cities, with its grand boulevards, magnificent palaces and museums, vibrant cultural activities and night life, and luxury shops. But most of its residents dwell in a world apart from the glitter of the *Paseo de la Reforma*. An increasing majority of them are forced to live in the miserable poverty and squalor of the conurbation's 500 slums, as well as the innumerable squatter shantytowns that form the burgeoning metropolitan fringe (the notorious *ciudades perdidas*, or lost cities). This is the domain of the newcomers, the peasant families who have abandoned the hard life of the countryside, lured to the urban giant in search of a better life. With Mexico's underemployment rate hovering above 30 percent in recent years, decent jobs and upward mobility quickly become elusive goals for most of the new arrivals. Yet despite the overwhelming odds, a surprising number of migrants eventually do enjoy some economic success.

in an informal way, urban places of various sizes. Hamlet, village, town, city, metropolis are all part of this hierarchy. At the top of these ranks is the largest, the *megalopolis*. A megalopolis is a coalescence of metropolises, a kind of urban region larger even than a conurbation. The largest one in the world lies here in the United States, extending from Boston and environs in the north to Washington and its suburbs in the south, thus including the New York-New Jersey conurbation as well as Philadelphia and Baltimore. Geographers named this urbanized region *Bosnywash*.

So in response to our question of what constitutes a city, the answer is a firm "that depends." Size is only one part of it. One country's town is another's city. The functions of the place matter more. Cities, large and small, perform services of many kinds, produce goods, provide leadership in government and culture. Cities influence and in some ways dominate the regions around them we call *hinterlands*, a German term meaning land behind the city. People in the hinterlands live their lives in the city's shadow, selling their products in the city (or having them shipped from there), banking there, using the medical facilities there, reading the newspapers published there.

Speaking of newspapers, they are often a good guide to a city's reach over its hinterland. Next time you're on a long road trip between major cities, say Kansas City and Denver or Chicago and Detroit, and you make a few stops along the way, see what city's newspaper sells in those gas station convenience stores. Somewhere along the interstate, the *Chicago Tribune* will disappear and the *Detroit Free Press* take over. Same with the *Denver Post* or the *Rocky Mountain News* and the *Kansas City Star*. Newspapers are one measure of success when cities compete for influence in their hinterlands.

A city and its surroundings, then, combine to form a large system of many interacting parts, some of which can be seen in the form of highways, railroads, power lines, telephone wires, dams, water pipes, and so on. Exactly where the city's boundaries are, though, is at times a problem. Cities expand, and their suburbs, in turn, gobble up rural land. This expansion leaves parcels of farmland actually surrounded by urban sprawl. Should this farmland, surely destined for development, already be counted as urban?

The people at the U.S. Bureau of the Census have particular trouble with this. So they, too, have various ways of defining a city, or, as the Census Bureau calls it, a Metropolitan Statistical Area (MSA). For example, an MSA is a city of more than 50,000 inhabitants—but an MSA is based on county units, so not only the actual city, but the whole county in which it lies is taken into account. MSAs, as a result, include a lot of farmland. Another Census definition is the Urbanized Area (UA), that is, the central city plus its immediate suburbs. And there are still other measures. So when you look up the population of your city in Census Bureau reports, be sure to note which measure is being used. City population figures can vary quite widely.

Here's an example: the *city* of Houston ranks fourth in size among U.S. cities, after New York, Los Angeles, and Chicago. But the Houston-Galveston-Brazoria Consolidated Metropolitan Statistical Area (CMSA) ranks tenth, after Dallas-Fort Worth and ahead of Miami-Fort Lauderdale. And speaking of Miami, the city doesn't even rank in the top 20, but its CMSA is eleventh in the county. Census figure readers beware!

Population data and census tracts may vary, but the lure of the city continues. True, inner-city problems have driven hundreds of thousands away from once thriving downtowns, but they haven't gone very far. If you read U.S. Census Bureau figures for our cities carefully, you'll see that the populations of what we might call the city proper have shown some decline. But this is more than compensated for by the mushrooming suburbs, many becoming so large that they're competing with the original downtowns for primacy in the overall urbanized area. And for all its troubles, the skyscrapered

RANKING CITIES VS. METROPOLITAN AREAS
IN THE UNITED STATES, 1990

Rank	City	Rank	Metropolitan Area
1	New York	1	New York-New Jersey-Long Island
2	Los Angeles	2	Los Angeles-Anaheim-Riverside
3	Chicago	3	Chicago-Gary-Lake County
4	Houston	4	San Francisco-Oakland-San Jose
5	Detroit	5	Philadelphia-Wilmington-Trenton
6	San Diego	6	Detroit-Ann Arbor
7	Detroit	7	Boston-Lawrence-Salem
8	Dallas	8	Washington, D.C., Area
9	Phoenix	9	Dallas-Fort Worth
10	San Antonio	10	Houston-Galveston-Brazoria
11	San Jose	11	Miami-Fort Lauderdale
12	Indianapolis	12	Atlanta Area
13	Baltimore	13	Cleveland-Akron-Lorain
14	San Francisco	14	Seattle-Tacoma
15	Jacksonville	15	San Diego Area
16	Columbus	16	Minneapolis-St. Paul
17	Milwaukee	17	St. Louis Area
18	Memphis	18	Baltimore Area
19	Washington, D.C.	19	Pittsburgh-Beaver Valley
20	Boston	20	Phoenix Area

Information from U.S. Census Bureau

downtown of the American city continues to symbolize the power, the energy, the creativity, the productivity, and the opportunity that long ago began to make New York, Boston, Chicago, Philadelphia, and Dallas the stuff of dreams.

ANATOMY OF THE CITY

Geographers like to look for regions. When we understand the regions, that is, the spatial components, of a continent, a country, or a city, we can comprehend how it all fits together and works. And cities, like other spatial systems, have regions. An ideal, or model, city is a bit like one of those toys made of different-colored rings that fit inside each other. There's the core, the downtown, or, to use the technical term, the Central Business District (CBD). This is where the land is most expensive, the skyscrapers the tallest, the sidewalks and streets the most crowded. Around this CBD lies a zone that shows, more than any other, the problems of the modern American city. This ring used to be called the Middle Zone, where you'd find blue-collar workers' houses, large-scale commercial enterprises (such as auto sales), and light industries. Today, this is the zone of urban decay and dysfunction, where poverty, crime, and other urban ills afflict the city most severely. Next comes a ring of newer industries and working-class houses, and beyond this zone we get into the suburbs, with ever larger lots and bigger houses.

On maps and aerial photographs you can still see this concentric city shape, but another pattern also reveals itself. In some ways, the city looks like a pie cut in uneven pieces, all pointing to the downtown. Highways and roads leading from the hinterland to the CBD help create this pie-shaped impression, but there's more to it than that. Sometimes, high-quality, high-priced housing continues from the most expensive suburbs all the way to the downtown's edge (as in the Chicago area along Lake Michigan on the city's north side and in Miami between U.S. 1 [Dixie Highway] and Biscayne Bay on the south side). Transportation systems (airport, railroads, highways) also create sectors penetrating all the way from the city's edge to the CBD, as do industrial zones in some cities. So whatever forces tend to shape concentricity in cities are countered by those forming sectors.

Today, American cities exhibit yet another spatial pattern. Ringed by growing, even mushrooming suburbs, downtowns are losing their dominance. Lots of people don't commute downtown to work, they commute from one suburb to another along one of our many beltways. Suburbs are acquiring their own downtowns and skylines, and the urbanized area looks like an amalgam of cities. My colleague Peter Muller, who teaches Urban Geography at the

University of Miami, has dubbed this the "pepperoni pizza model" of the modern American city. And each of the downtowns in the urbanized area has its own rings and sectors—no wonder the American city is a spatial labyrinth!

REVIVING THE DOWNTOWN

If the encircling inner cities have their troubles, the urban cores, the downtowns, suffer, too. The emigration to the suburbs is a large part of the problem. As well-to-do taxpayers moved from their central-city condominiums and mansions to the distant suburbs, the city lost their tax revenues, and schools, services, maintenance, police, fire protection, waste removal, and other tax-supported operations had to be pared down. That, in turn, made the central city a less attractive place to live, and the vicious cycle continued. Soon, census data based on city-proper municipalities began to show significant declines in their populations. And the big-city mayors appealed to the federal government for help in their battles to keep their cities viable.

We Americans don't like to be constrained. Our cities and their sprawling suburbs use space undreamed of in Bombay, Beijing, or Buenos Aires. We drive many miles to and from work on gasoline that is the cheapest by far in the industrialized world. A free standing house on a spacious lot became the goal of millions of families, and the suburbs sprang up, away from the woes of the central cities.

But what of the downtowns, sites of great museums, libraries, concert halls, monuments, churches, and architectural treasures? Corporations could move away, and many did, to office and industrial parks in the suburbs. But you can't as easily move great universities, historic squares, old cathedrals, or venerable theaters. Shouldn't America's downtowns be supported as part of our cultural heritage?

Not everyone would answer this question affirmatively. If the American downtown has served its purpose and no longer functions

SUBURBAN DOWNTOWNS

Mushrooming "suburban downtowns" have transformed the U.S. metropolitan landscape since 1980, and these edge cities are now beginning to appear in Western Europe, too. The best known and most important suburban business complex in Europe is *La Défense*, located just west of Paris, whose growth has been encouraged over the past two decades in order to relieve congestion in the French capital's central business district.

By the early 1990s, it had become evident that the planners had carried out their mission only too well: this burgeoning activity has not only achieved unintended primacy in its own metropolis, but it has also become continental Europe's largest business district, with total annual transactions exceeding the gross national product of the Netherlands. Among other things, La Défense is now home to Europe's largest concentration of computer-age office space, its biggest shopping mall, more than 150 corporate headquarters facilities, the European head offices of dozens of prestigious multinational corporations, a massive hotel-convention-restaurant-entertainment complex, and a workforce of more than 135,000 people (more than half of them in management positions).

Perhaps the most striking aspect of La Défense is that it is so much more than just another large-scale suburban downtown. Its centerpiece, the enormous Grande Arche, completed in the early 1980s, anchors the west end of Paris's Historic Axis, which now stretches over five miles (eight kilometers) from the Louvre up the Champs d'Elysées to the Arc de Triomphe and then outward to La Défense. This architectural jewel, surrounded by the more than 50 unique sculptures that line its vast esplanade, has already become one of the most famous Parisian monuments, attracting an astonishing 26 million visitors each year.

Despite those numbers, the regional planners of La Défense have designed a new transport network that is proving capable of handling the huge traffic flow. Besides five expressway connections and behemoth parking garages for automobiles, five separate rail systems converge here—regular Metro (subway) lines, a new express Metro line, a commuter railroad line, a long-distance passenger-rail line, and a new high-speed line that will open in mid-decade. This represents some of the most advanced transportation planning in the world, and the thriving success of La Défense will have important lessons to teach the city builders of the twenty-first century.

as it once did, it should be left to its own devices, just as the industries of the Rustbelt and the communities in the urban Middle Zones were when their turn came—so argued many a columnist and television commentator. But others saw it differently. This, they said, is the time to come to the aid of what is in effect an American icon, a defining component of this society's cultural landscape. Government and business should cooperate in reversing the fortunes of America's central cities.

We have all seen the successes and failures. St. Louis was bequeathed a symbolic arch and, like Cincinnati and New Orleans, a downtown sports facility. Detroit's huge waterfront Renaissance Center was to revitalize a CBD that will need far more help than it can possibly provide. Copycat inner-harbor shopping mall-theme park developments arose from Miami to Boston. City streets were closed and turned into pedestrian shopping malls to compete with the suburban malls that draw customers from the downtown's once trendy avenues. But none of this has yet generated a full-fledged return from the suburbs. It will take much more.

The following tale of two cities—New York and Chicago—explains why. Within three weeks of my wide-eyed arrival in the New York City of 1956, I walked countless miles—from the United Nations to the West Side's waterfront, from Wall Street to Central Park. I marvelled at the crowds, the bustle, the noise, and the energy, but also at the orderliness, comparative cleanliness, and openness of the city. There was so much to do, and so much of it was free. Above all, there was such variety, in neighborhoods, street scenes, atmosphere, languages, foods. It was a city that never slept, whose streets were filled with activity even at three o'clock in the morning.

Later, during my days at Northwestern University, I came to know, and appreciate, Chicago. Northwestern is located in Evanston, north of the Loop. Suddenly I was a suburbanite, living in the attic of a stately residence owned by famed geography professor W. H. Haas. I rode the El (for elevated train) back and forth to the city countless times, enjoying the fabulous Chicago Art Museum, the great Chicago Symphony, then under the baton of Fritz Reiner, and those lovable losers, the Chicago Cubs, whose hero was Ernie Banks (I've been a Cubs fan ever since). But the Loop was no Manhattan. Chicago's downtown, late at night, became a one-street city: State Street. Even Michigan Avenue quieted down. Over the years, the action moved north, where Mr.

Kelly's was the great jazz club. Downtown Chicago increasingly became a commuter's, and decreasingly a residential, area.

And that, it seemed to me, was the crux of the matter. Those glass towers that rose over the cityscape—not only of Chicago but of so many other cities as well—may create impressive skylines, but they do not necessarily do anything for the ambience of the downtown. There should have been two rules: first, create an open, accessible atmosphere at sidewalk level, not a forbidding wall of banker's gray glass. And second, if you're going to build tall buildings downtown, then part of the space should be devoted to residential use, as a matter of zoning regulation. Given the office-space glut that has prevailed in city centers throughout America since the late 1980s, many builders undoubtedly wish that they had mixed their occupancies this way.

Where such combinations have been tried, they have worked quite well. In Washington, D.C., there is an increasing number of business-residence combinations downtown, and the local press has reported on their helpful contribution to the well-being and livability of the urban core. In Chicago, come central-city residential high-rise projects have been successful, too, notably the landmark twin Marina Towers on the river, now aging but instrumental in reminding people that the city can be a good place to live as well as work.

WHAT'S SO GREAT ABOUT SUBURBS?

I realize that suburbia is as American as motherhood and apple pie, and to question the pleasures of suburban living is almost as provocative as voicing doubts about greenhouse warming: you're up against a deep belief. The American dream today is the single-family house on a roomy lot with a two-car garage and two cars in it; a suburban, office-park workplace; a nearby shopping mall with all the routine requirements; and local government to protect against big-city interference. And don't disturb the dreamer with arguments about regional, beyond-the-suburb needs: water supply, waste disposal, traffic amelioration, law enforcement, retirees and other low-income people, and so on. The suburban ideal, in one form or another, is now reality for so many millions of Americans that a great suburban tide has transformed much of the countryside. You can marvel at its impact every time you fly into any of our cities. Many geographers are enthusiastic proponents of suburbia. I'm not one of them. Suburban life may be celebrated as an escape from the city, but in this process of dispersal, community life in urban settings largely went by the board. For centuries, indeed for millennia, we had learned to live in close proximity. Even during our cities' streetcar era, walking was a major part of daily routines. Sidewalks were the ribbons of activity that connected people and places, and they promoted interaction.

All this was left behind in a few short years. Front lawns replaced sidewalks, driveways replaced front doors as functional outlets. The automobile came to dominate lifestyles. Workers began to spend hours commuting—hours that should be spent with families. Youngsters depended on wheels to meet their peers, and many grew up bored and disconnected. Shopping became a car trip to the mall, not an interactive walk down main street. Oldsters no longer able to drive found themselves isolated in their houses or forced into nursing homes; in the city, their useful lives would have been much longer. Bars disgorged patrons who would have caused no danger walking home, but who killed countless people while driving drunk. In an ultimate misconception, some observers described the city-splitting four-lane highway as the new Main Street U.S.A.

Suburban life has its positive aspects: privacy, comparatively clean air, lawns and gardens, bugs, barbecues. But this space and resource-consuming phenomenon, for all its current popularity, is still a new experience. We have built suburbia, but we have yet to invest it with a sense of community—community within the suburb as well as beyond, where the enabling infrastructure lies.

WHY ARE CITIES WHERE THEY ARE?

Now here's an obvious geographic question. When you look at a map, you can see that, in *most* parts of the world, the largest cities lie farthest away from each other; smaller cities lie closer together; towns lie at still shorter distances, and villages and hamlets lie closest of all. Take a road trip across the Midwest, and you pass villages with regularity, every 15 or 20 miles (24 or 32 kilometers) or so, towns maybe every two or three hours—but to drive from one major city to another can take all day or more.

Of course, there are always exceptions. Along the East Coast from Boston to Washington, in the North American megalopolis, you go from city to city in an hour or less; you seem to be in Washington before you've left Baltimore. In the area of the Great Lakes, in what is called the Chicago-Pittsburgh corridor, *Chipitts*, cities also lie closer than average. But it's a long way from Chicago to Minneapolis, from Minneapolis to Denver, from Denver to Kansas City, and from Kansas City to Dallas. In most of the world (maps of South America, Africa, India, China will confirm this), the largest cities tend to lie well separated. Clustering of large cities occurs (in Japan, for example), but it's unusual.

The *where* question really has two answers. Many cities that thrive today were originally established because of what geographers call *site* advantage. There were lots of settlements along France's Seine River a couple of thousand years ago, but one of them lay at the place where an island split the water into two relatively narrow channels. That made this a favorite place to cross the Seine, and the island was conveniently secure, protected by natural moats. Soon, there were sizeable settlements on the island, on the north bank, and on the south bank—and Paris was born.

So Paris got a head start, but eventually the island (still called Island of the City, *Ile de la Cité*) ceased to have any real importance as a crossing point or as a fortification. You could cross the Seine in many places, and Paris was just one of them. But by that time, Paris had something else: *situation*. At the time the original settlement formed, its founders did not know it, but they chose a place with excellent relative location, at the heart of what we now call the Paris Basin. Farm produce flowed in from all around, markets thrived, workshops prospered. Soon the powers of the state and the church focused here (the great Notre Dame Cathedral was built on the Ile de la Cité) and Paris's primacy was assured. Paris has had no competition in France since. With nearly 9 million inhabitants, the city accounts for one-sixth of the population of all of France!

This is the story of many great cities: their sites were desirable (as ports, river crossings, gateways, defenses), and later their situations proved advantageous over others. There was a string of settlements from Maine to Miami along the East Coast, but only a few of them evolved into a great metropolis. And of the tens of thousands of interior settlements, we now know only such places as St. Louis, Kansas City, Denver, and the like as major cities.

Earlier this century, a geographer named Walter Christaller tried to do what scientists like to do: to figure out a way to predict which cities will thrive and which will not. He was one of geography's first model builders, and he theorized that, all things being equal, an urban system would develop with hexagonal symmetry. If you put two settlements in such a model and give them each circular hinterlands, then their hinterlands would overlap. Draw a line to divide that overlap, and do it in all directions, and you'll draw a hexagon around each settlement. Assume now that settlements will form in the most remote parts of these hinterlands, where the meeting points of the hexagon are located, and that each of these smaller settlements also has its own hexagonal hinterland. Soon the map looks like a network of interlocking hexagons, as in the diagram. At the center of the largest hexagons lie the major cities. So if all the land was flat and there were no coasts or rivers or mountains to interfere with all this, Denver and Kansas City and Minneapolis-St. Paul would lie at the centers of great hexagonal hinterlands contain-

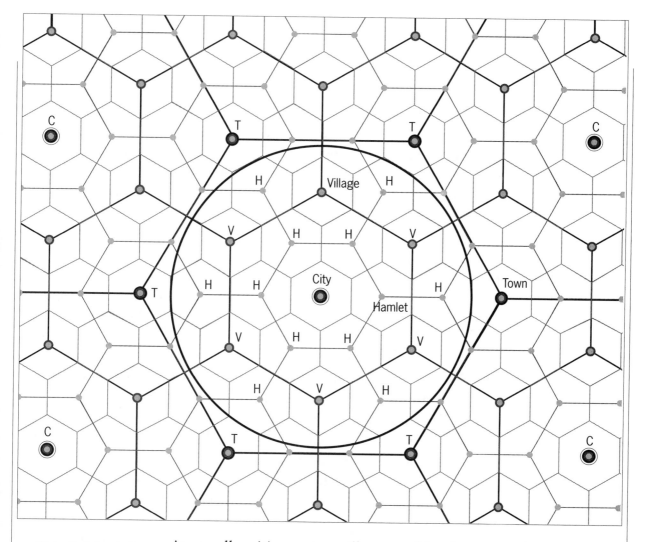

Christaller's interlocking model of a hierarchy of settlements and their service areas. H = Hamlet; V = Village; T = Town; C = City.

ing smaller cities, towns, villages, and hamlets at regular intervals.

Christaller set in motion a field of geography called location theory. Today, it's all mathematics and computers, but no shopping center or new airport or other facility is planned without reference to it. Today this field is replete with mathematical formulas and computer-generated models. Christaller's basic assumptions, the ones he used to derive his hexagonal framework, almost never apply: the world is full of mountains and rivers and coastlines and other kinds of interference in the system. Such exceptions help explain why several metropolises lie close together in the megalopolis of the Atlantic Seaboard and the Great Lakes area. But where the land is generally flat and interruptions are few, the pattern Christaller envisaged does come close to reality. One especially interesting study was done on the great flat North China Plain, where tens of millions of Chinese live in villages and move, predominantly, on foot or by bicycle. Rivers and irrigation channels are about all that interrupt the landscape; and as a result, the distribution of villages and towns comes very close to the Christaller model.

WORKING CITIES

Cities and towns are sometimes dominated by one particular activity—or they owe their origins to such an activity. We all know of Colorado's mining towns, of Florida's retirement communities, of Arizona's resort towns, of New England's fishing ports, of Iowa's farming centers. Mention Las Vegas, Hollywood, or Sun City, and we have no doubt about the main function of each.

This notion of the original (or current) function of a town or city can serve as an interesting geographic discovery project. One of my distinguished colleagues at the University of Chicago, Chauncy Harris, started all this by making a series of maps that showed how functional cities are distributed in our country. When cities get larger, their original function tends to get overtaken by diversification, so that many larger urban centers are multifunctional. But others remain dominated by one function. There are hundreds of college and university towns in this country, for example, that would wither on the vine if their educational function ended.

Professor Harris started the ball rolling in 1942, using several categories and showing, for example, how strongly manufacturing centers were concentrated in the East and Northeast, retail centers in the heart of the country, and so on. When you remake his maps based on the 1990 Census, you can immediately tell how much has changed: manufacturing has diffused westward, retailers and other services are now much more numerous and in more cities, and so on. The United States is one of the most urbanized societies in the world, and cities have had to change functions with the times. But you can still see specialization on our urban map, as a little research in your home state will quickly prove.

Some of us undoubtedly got our first taste of geography (or, rather, of geographic trivia) this way: by learning the names of the capitals of the States. Many of these places, too, are functionally specialized: they depend wholly or largely on the activities arising from the State government. If you make a map of this, you'll find that quite a few such cities have two main functions: government and education. Tallahassee, Florida, is the seat of the State government and also the home of a fine university with a good geography department, Florida State University. Madison, Wisconsin, and Columbus, Ohio, are among quite a few other government-university towns.

When foreigners learn about the capitals of our country, it is often something of a surprise. In one of my exams at school in Holland, I listed San Francisco as the capital of California. When my teacher said that I had it wrong, I said that it must be Los Angeles. Sacramento? I had never even *heard* of Sacramento. (These days, I gather, many Californians also wish they'd never heard of Sacramento.)

Unlike many, though not all, other countries, the idea here was to make smaller cities State capitals so that government would not be submerged by the action in the bigger cities, and to put the capitals somewhere nearer the geographic centers of the State so that they would be in touch with the heart of the territory. Lansing, Michigan; Springfield, Illinois; Harrisburg, Pennsylvania; and Albany, New York, for example, are not exactly the leading metropolises of their States. There are the exceptions to prove the rule: Salt Lake City, Utah; Atlanta, Georgia; Denver, Colorado. Still, the great majority of American State capitals are modest-sized towns in which the government's facilities and payroll are crucial to the local economy.

In my television series on *Good Morning America*, I've talked about the prospects of individual cities, such as Mexico City, and about broader issues, such as the uneven quality of education in our urban areas and the changing commuting patterns we mobile Americans are creating. Often, my mail includes requests for additional information about the geography of our (and the world's) cities. I can think of no better or more interesting book than *Interpreting the City: An Urban Geography* by my colleague Truman Hartshorn, who is Professor of Geography at Georgia State University. It's the story of urbanization and a tale of many cities, and it provides answers to countless questions this brief chapter could not even touch upon.

C H A P T E R   6

CAPITAL GEOGRAPHY

When I visited Moscow for the first time, in 1964, I entered Red Square early one morning and saw tens of thousands of people in a seemingly endless, winding line waiting to see the embalmed body in the Lenin Mausoleum. I learned that that line never stopped forming, day after day, year after year. People would wait in it for as long as 10 hours. I realized that this line was a kind of ongoing national demonstration, the sort that makes a capital city what it is.

Just before my part of the line reached the entrance, we neared the Kremlin wall where the graves of national figures were marked by impressive busts atop marble columns. Josef Stalin's grave, though, was marked only by a small black marble plaque in the grass at the foot of the wall. Just as a Russian next to me pointed to it, saying, "Stalin, Stalin!" an old woman suddenly appeared, scurrying quickly along the base of the wall. She wore a white scarf over her head and was bent over, carrying, it seemed, something in her blue apron. When she reached Stalin's grave, she opened her apron and out fell a huge bouquet of flowers. Then she ran and was quickly lost in the Red Square crowd. A uniformed guard from the group guarding the Lenin Mausoleum's entrance hurriedly picked up the flowers and carried them away. I turned to see the reactions among the Russians near me in the line. Every single one of them was looking in the opposite direction.

Many of the world's great cities are more than major metropolises: they're the headquarters of the state, focus of the nation, repository of history, altar of ideology, symbol of strength. Throughout history, nations have invested heavily, sometimes excessively, to give their capitals an image of wealth and well-being, to make them mirrors of the national culture. Athens, Rome, Paris, Beijing, Tokyo, Washington, Lima, Bangkok—all carry the impress of national aspirations.

And it works. Richly endowed capital cities from London to Kuala Lumpur are magnets for visitors from home as well as abroad. All those impressive, columned buildings, religious shrines, museums, monuments, statues, and national cemeteries not only draw crowds, they help cement the nation's bonds and instill a sense of purpose. Foreign tourists, seeing the lines of local people waiting to enter the memorial of a national hero, may realize for the first time how strongly people feel about their country.

Capital cities are used to educate, even to indoctrinate. For decades, a huge portrait of Mao Zedong hung over the entrance of Beijing's Forbidden City, overlooking fateful Tienanmen Square. You couldn't miss it from the farthest corners of this, the world's largest public square; you had to walk beneath it when you entered the Forbidden City, former domain of China's dynastic rulers. What better place to remind one and all of Mao's thoughts?

PRIMATE CITY

Obviously, a country's capital city is more than just another urban center. Often, though not always, it is also the largest city, as in the case of Athens, Paris, London, Buenos Aires, and Mexico City. A noted geographer, Mark Jefferson, made an exhaustive survey of such capitals and of the largest cities in countries where capitals are *not* the largest, as in the United States, and in 1939 proposed the Law of the Primate City. "A country's leading city," he wrote, "is always disproportionately large and exceptionally expressive of national capacity and feeling."

That was still true around midcentury, when most capitals were the centers of older countries, many of them colonial powers. But when colonies became independent, their largest cities more often than not became capitals—but these were hardly expressive of "national capacity and feeling." Rather, they were reminders of the bygone colonial age.

Many former colonies tried hard to create, in short order, the atmosphere Jefferson's "law" implied, often at huge cost. Endowing Accra with such new adornments helped drive Ghana into bankruptcy; Addis Ababa became, in the words of one geographer, "a shield behind which the *real* Ethiopia hides." (True, Ethiopia was never colonized, but it was overrun by Italy and emerged from the Second World War determined to match other African countries in the splendor of their capitals.) But what makes capitals special places is the slow, accretion of a nation's tangible history. Rome, as they say, was not built in a day.

STATES, NATIONS, AND CAPITALS

It is sometimes frustrating to hear geographic terms misused in ordinary conversation, often leading to confusion about the topic at hand. I could make a long list, but near the top of it would be the words "state" and "nation." This is especially relevant in the context of capital cities.

The correct term for an independent, functioning country is a state. (Some remaining countries are not independent, for example, Bermuda; some are not functioning, such as Somalia.) When the great majority of the people in such a state belong to the same nation, for example, in Greece, Ireland, or Chile, that country is often referred to as a nation-state. But in many states there is not just one nation, but two or more. Belgium, for example, is a state of two nations: Flemish and Walloon. So are Cyprus (Greek and Turkish), Sri Lanka (Sinhalese and Tamil), and, until their recent "velvet divorce," Czechoslovakia (Czech and Slovak).

But what is a nation? That is a difficult, often sensitive question. In South Africa, polls confirm that the great majority of the more than 8 million Zulus consider themselves to be a nation within the state. In these turbulent times, the sense of nationhood is strong in multinational states such as Russia, Canada, Nigeria, India, and the former Yugoslavia. At times, such nationalism threatens to break the state apart, as has happened in Yugoslavia, violently, and in Czecho-slovakia, peacefully. The map of the states of the world is not the same as the map of the nations of the world. People require a shared language, religion, history, and tradition, and actual or perceived discrimination to galvanize and sustain feelings of nationhood.

In states with one dominant nation, that is, in nation-states, such divisiveness is not a problem. State as well as nation is focused on an often old or even ancient capital city that is, indeed, the primate city. Athens, Rome, Madrid, Paris, London, Warsaw, Cairo, Bangkok, Beijing, Manila: these are the capitals of centuries, if not millennia. And the nation-states they represent can afford to have centralized governments.

In states where the feelings of more than one nation must be accommodated, both the state system and the capital are likely to reflect it—and if they don't, the state may not survive. In the former Czechoslovakia, not only did the Slovaks feel themselves to be the disadvantaged nation compared to the Czechs, but many felt that Prague was the Czech, not the Czechoslovakian capital. In Belgium, the capital city, Brussels, does not seem to rank highly in the emotions of Flemish *or* Walloon.

When such problems arise, there are two obvious solutions. One is to establish more than one capital and to divide the various functions of capital cities among these. Several countries have or have had multiple capitals: South Africa (Pretoria and Cape Town), Bolivia (La Paz and Sucre), Laos (Vientiane and Luang Prabang), Libya (Tripoli and Benghazi), and, for rather different reasons, the Netherlands (The Hague and Amsterdam). Such arrangements reflect not only cultural differences in these states but also religious and royal imperatives.

The other solution is to create a compromise capital, a kind of neutral headquarters that does not represent one national group in a state over another. Americans are familiar with this option: Washington, D.C., was established on land ceded to the government by Maryland and Virginia. This way, neither New York nor Philadelphia would become the seat of government to the disadvantage of the other. The United States is a federal state, and its capital remains (Statehood proponents and shadow senators notwithstanding) a federal territory. Other countries, several of them also federal states, have adopted the same solution.

One of the most interesting experiments of this kind occurred in Australia, where the two great cities of Sydney and Melbourne (themselves State capitals of New South Wales and Victoria, respectively) vied for the role of national government headquarters. Shortly before the Australian federation was established, nearly a century ago, the premier of New South Wales managed to secure an agreement that a new federal capital would lie within his State—in return for two concessions. First, the capital would *not* be Sydney, but a totally new capital to be built on a federal territory to be jointly delineated. And second, Melbourne would be allowed to function as the temporary Australian capital until the federal headquarters were ready for occupation.

The site of the Australian capital, Canberra, was chosen in 1908. A worldwide architectural design competition was held to secure the best possible city plan, a competition won by the American architect-planner Walter Burley Griffin. In 1927, the city, with impressive government buildings, wide avenues, lakes, and fountains, all overlooked by Mount Ainslie, was ready for occupation. Some 940 square miles (2,400 square kilometers) of land were carved from New South Wales and made the Australian Capital Territory, and today Canberra, also the site of a major university, is the only large Australian city that does not lie on the coast.

So what kind of a city is Canberra today? Forgive me, but I find it about as exciting as the inside of a freezer. True, it's the seat of power, the scene of political contest. From various vantage points you can see the grand design. But to me (and, I'm told, to many Australians) this is just too antiseptic a place to live in for very long. There's none of the action of Sydney or the culture of Melbourne—well, very little of it anyway. Artificial, introduced cities take a very long time to develop character—if they ever do. My first visit to Canberra reminded me, actually, of an earlier excursion to another federal capital, Brasilia in Brazil. I'll never forget this: I arrived on a Thursday, spent the night, and went for a drive the next day. What I saw was a mass emigration. Dozens of airplanes ferried thousands of people from Brasilia to Rio de Janeiro for the weekend; on Saturday and Sunday, Brasilia was a ghost town. Sunday evening and Monday morning, the airport was a beehive again, planes disgorging the Rio weekenders for the coming workweek.

The geographic story of Brasilia is a bit different from Canberra's. In Brazil there was no big-city competition (although there were people in São Paulo who would have liked to be given a chance). In fact, Rio de Janeiro was Brazil's capital and primate city for many years. After the Second World War, Brazil's (and Rio's) explosive growth created congestion and inefficiency; the government also saw a need to get Brazilians away from their coastal orientation. Even today, the vast majority of Brazilians live within 200 miles (322 kilometers) of the Atlantic coast. A new, interior capital city would have two desired effects: it would relieve Rio of its clogged governmental functions, and it would create what geographers call a *growth pole* in the distant interior. This, it was hoped, would convince Brazilians to look toward the great western frontier that lay open to penetration and development.

On the principle that Brazil had many fine architects, a national (rather than worldwide) competition was held to find the best design for Brasilia. The winning layout resembled a giant airplane, its wings spread roughly north-south and its nose pointed symbolically toward the Brazilian interior. In 1960, the capital functions were transferred amid expressions of hope that that year would usher in a new era of westward expansion.

Today, Brasilia's population approaches 1 million, and the more informal areas of the place have come to look like a "real" city. Canberra's population is only about one-third of Brasilia's, but consider this: Australia has a mere 15 million inhabitants, Brazil 10 times that many. If growth had been proportional, Brasilia today would have had over 3 million inhabitants! So once again, the planned, artificial, synthetic city has a hard time establishing an urban soul, even in multicultural Brazil. Christaller would have known it all along.

CAPITAL AS WEAPON

What the Brazilians did was to use their capital to attain a national objective: to refocus Brazilians' attention toward the unopened interior, where, as the newspapers then and now predict, lies the destiny of the state. Thus Brasilia became a beacon as well as an administrative center, 700 miles (1,100 kilometers) into the *mato grosso*.

That is a relatively benign (and domestic) goal. Other countries, however, have relocated their capitals not so much to reorient their own populations as to make a geopolitical statement to the outside world.

Political geographers call such capitals *forward capitals*, signifying their placement in peripheral, sometimes contested areas. When the Germans decided on Berlin as their capital originally, it was clear to Europe and to the rest of the world that German intentions toward Eastern Europe were serious. Berlin lay close to the German-Slavic cultural frontier, far to the east of the center of gravity of the German state. Now, Berlin lies a few dozen miles from the Polish border, so when the reunited Germans decided to move their capital from Bonn, which had served West Germany, to their old headquarters, there were those who heard the echo of past campaigns.

A different example occurs in Pakistan, where coastal Karachi, the former colonial headquarters, at first served as the capital. But the Pakistanis have long had quarrels with their neighbors over the country's far north, including Jammu and Kashmir (where conflict with India is chronic). And so Islamic Pakistan decided to make a geopolitical statement with its new capital: it repositioned the government from southern Karachi to northern Islamabad, on the very doorstep of its embattled frontier. This left no adversary in any doubt: the state's presence in these remote highlands is as powerful as it is on the coast or in its core area, the populous Punjab. With this action, Pakistan used its capital as a weapon, to warn its neighbors of its intentions.

Capitals, then, are more than meets the eye. There's a geographic story attached to almost every one of them. Too bad that so many trivia questions ask only about the name of a country's capital and not about the often fascinating story that lies behind its location. For example:

Burkina Faso.   You may not have heard much about this landlocked West African country, but its capital's name is a favorite with trivia buffs: Ouagadougou. The French annointed this place, at the end of the railroad from coastal Abidjan, as the centrally located headquarters of the colony they called Upper Volta.

Indonesia. Its capital, Jakarta, dates from 1527, when a Muslim prince defeated Portuguese invaders here at the mouth of the Ciliwung River, northwest Java. In the local language, Jayakarta means Great Victory, and the place has carried this name nearly five centuries—except during the Dutch colonial occupation, when it was called Batavia.

Japan. Has had three capitals: Nara, the original center; Kyoto, city of magnificent religious shrines and serene gardens; and, since the late 1860s, coastal Tokyo, formerly the fishing port of Edo. When Japan's modernizers took control of the country during the Meiji Restoration of 1868, they made Tokyo the symbol of the new Japan: outward-looking, commercial, competitive, expansionist. Soon Tokyo was the center of a colonial empire.

Kazakhstan. The capital of this former Soviet Central Asian republic, Almaty (formerly Alma-Ata), lies in the far southeast corner of this Russian-Kazakh country, and thus in the Muslim

Kazakh sector. Now and then there is talk of moving the government to a city in the Russian (northern) zone, but it won't happen. Ethnically fragmented Kazakhstan has enough problems not to add capital relocation to them.

Liberia. Named its capital, Monrovia, after an American President: James Monroe.

Mexico. May soon have the world's largest capital—and largest city. While the growth of the Tokyo urban area slows down, that of Mexico City still speeds up.

Mongolia. Must have one of the world's most godforsaken capitals, in Ulaan Bataar. The place happens to have a sister city—Laramie, Wyoming.

Netherlands. Has two capitals, but the seat of government lies in The Hague, not Amsterdam. Technically, since the sovereign is crowned in Amsterdam and the crown has ultimate sovereignty, Amsterdam qualifies ceremonially—but the action is in The Hague.

New Zealand. When a country consists of two major (and a few smaller) islands, the location of the capital poses a problem. New Zealand's capital is Wellington, which is on the North Island directly across the narrowest part of the Cook Strait from the South Island. Earthquake-prone Wellington lies right on the major geologic faults marking the contact between the Pacific and Australian tectonic plates. Fasten your seat belt!

Nigeria. Is in the process of relocating its capital from the old colonial center, Lagos, to Abuja, near the confluence of the country's two major rivers, the Niger and the Benue. Nigeria is a federal state, and Abuja is situated nearer the country's geographic center *and* in its transition zone from the Muslim North to the Christian-Animist South. But the grandiose scheme, begun when oil riches poured into Nigerian coffers, has been scaled back.

Tanzania. Can't seem to make up its mind about its capital. When the country became independent, it was Dar es Salaam (haven of peace) on the Indian Ocean coast. Then it was decided to move the government more toward the country's interior, to a place called Dodoma. But not all government offices moved, and now Dar es Salaam seems to be the effective capital again.

Thailand. Has what may be the most environmentally degraded capital in the world today. Its air makes Mexico City's seem clean: it is said that breathing Bangkok's air is the equivalent of smoking five packs of cigarettes a day. The waters of the Chao Phraya and the canals that lead into it are filled with garbage. *And* the land on which the city is built is sinking!

United States. Now here's a twist—a capital, Washington, D.C., that wants to be a State! While other federations carefully separate their capital cities from State jurisdictions by establishing federal territories for them, politics in America are enlivened by a campaign to make a State out of just such a federal entity. It isn't likely to happen. The justification for a federal capital in the District of Columbia (D.C.) is every bit as strong as that for Brasilia, Canberra, Abjua, or New Delhi. Washington's municipal government has not given evidence of its capacity to effectively run a city, let alone a State. Among the proposed names for this sliver of territory is New Columbia.

CHAPTER 7

BOUNDARIES: NEIGHBORS GOOD AND BAD

We all live in what geographers call bounded space—in countries, provinces, districts, private properties, and other legally circumscribed territories set off by boundaries. These boundaries affect us every day, in the passports we carry, the taxes we pay, the schools we can and cannot attend, the votes we cast, the homes we buy. Robert Ardrey, in his fascinating study of our need to divide and subdivide our activity space, called it "the territorial imperative," a trait we share with many animals.

Boundaries have become unavoidable; they are necessities in our complex and competitive world. But we have learned to our sorrow how divisive and disruptive they can be. That is why many states and nations today are trying to overcome the barriers between them. European countries want to enable people, goods, and money to cross their common borders with minimal impediment. The North American Free Trade Agreement (NAFTA) was designed to lower economic obstacles between the United States and its two neighbors, obstacles symbolized by the political boundaries between Canada, the United States, and Mexico.

Even as efforts of this kind proceed, however, new boundaries are being drawn. International borders now separate the Czech Republic from Slovakia, Slovenia from Croatia, Eritrea from Ethiopia, Ukraine from Russia, Turkish northern Cyprus from the Greek south. All these international boundaries, and many more, have been created (or elevated in status) in just the past few years.

On the other hand, some divisive boundaries have been overcome. The Berlin Wall and the boundary between West and East Germany have been swept away. Vietnam was unified. There now is a glimmer of hope that the boundary that separates the Korean nation into two ideologically opposed states may eventually go the way of the Iron Curtain.

A LINE IN THE SAND?

During the 1991 Gulf War, the inviolability of international boundaries was a big issue. Iraq occupied Kuwait with the intent of annexing it as its nineteenth province. Kuwait, and its United Nations allies, insisted that the border between Iraq and the sheikdom was internationally recognized and permanent. Generals, news reporters, journalists, and others often referred to the "line in the sand" beyond which Iraq should not be allowed to go, a line that marked the limit of national jurisdiction.

International political boundaries, however, are not merely lines in the sand, on the ground, or anywhere else, for that matter. True, they appear on atlas maps as lines, but that's because atlas and other sheet maps are two-dimensional. Boundaries must do much more than mark off national, provincial or county territories on the surface. Boundaries also divide what's below the surface and even what's above it. In fact, boundaries are vertical planes, imaginary curtains that separate airspace above the ground and resources below it. Thus that line on the map (and in the sand) shows only where the vertical plane of the boundary intersects the topography of the Earth.

DIVIDING RESOURCES

It is one thing to mark a boundary on the surface, but it's quite another to locate it deep below the ground. Take a look at the Dutch province of Limburg, where the town of Maastricht is located (now famous because that's where the European Community agreements were drawn up). Limburg is a narrow corridor of a province, wedged between Germany to the east and Belgium to the west. Underneath lie rich coal seams that extend from Belgium through Limburg into Germany.

The Dutch mined these coalfields vigorously. They got down to the seam, and then excavated it eastward. It wasn't long before they were mining coal below Germany's terrain, but of course there was nothing down there to tell the miners that they were crossing the

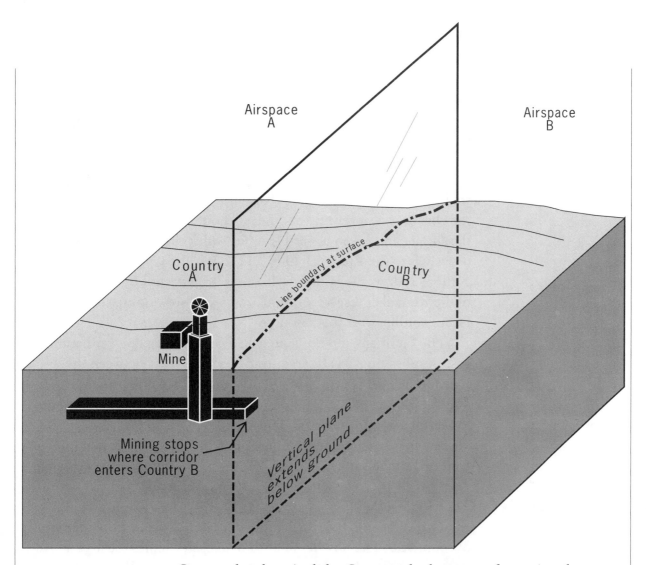

Airspace
A

Airspace
B

Country
A

Line boundary at surface

Country
B

Mine

Mining stops
where corridor
enters Country B

Vertical plane
extends
below ground

A political boundary is a vertical plane, not merely a line on the ground.

German border. And the Germans had no way of stopping the Dutch underground; they were mining the same seams, but elsewhere. Soon there was some industrial espionage, the Germans got hold of a map of the layout of the Dutch coal mines and claimed violations of their boundary. A pretty good quarrel followed, and the Dutch agreed to stop mining eastward—but only after they had secured a substantial piece of the German reserve.

Things can get even more complicated when it comes to oil and natural gas. Many oil fields and gas reserves extend from one country to another beneath their common boundary. So it was, in fact, where Iraq and Kuwait meet: the Rumailah Oil Field extends from northern Kuwait into southern Iraq. Here's the problem: when one of the neighboring countries starts to draw on such a joint reserve, oil or gas will flow from the subsurface of the nonexploiter to that of the exploiter. The Iraqis accused the Kuwaitis of drilling oblique wells, so the pipes standing in Kuwait actually fed on oil below Iraq; the charge was never proven. But without doubt, oil flowed below the surface from Iraq toward Kuwait, which drilled much more vigorously into the Rumailah Oil Field. After the Gulf War, the United Nations redrew the boundary between Kuwait and Iraq, putting all of the Rumailah reserve within Kuwait and preventing, so it hoped, a future conflict over this resource.

The Germans and the Dutch also were at it again over natural gas. A substantial gas reserve straddles the border between the Dutch northern province of Drenthe and adjacent Germany. Massive Dutch exploitation of this underground balloon of gas led to German complaints—but how could the loss to Germany be quantified? Contentious boundary issues, obviously, are not confined to the surface.

ASSIGNING BLAME

Above the ground, too, boundaries are used to protect the interests of countries. The now-familiar term *airspace* defines the area above a state's territory where the rights of other states may be limited: for example, commercial airplanes from particular countries can overfly other countries only by official, and often reciprocal, agreement. During the *apartheid* period in South Africa, other African countries prohibited South African Airways from flying over their territories, forcing SAA into long, expensive oceanic routes. During the communist period, the Soviets established strict corridors for commercial aviation; when Korean Air Lines 007 strayed over Soviet territory where such corridors did not exist, it was shot down with all aboard.

There was no retaliation: the Soviets had exercised their right to protect their airspace within their boundaries.

The international boundary curtain above the ground produces other issues. Imagine this: country A develops a major industrial complex. Clusters of factories consume coal, iron, and other raw materials and produce large quantities of products ranging from toys to heavy equipment. Not much care is taken to protect the natural environment, and a forest of smokestacks belches chemical pollutants into the air.

A large part of neighboring country B lies in the path of prevailing winds, and more often than not, heavily polluted air from country A wafts over country B. When it rains, the precipitation becomes a shower of acid, and plants, forests, water supplies, and other assets of country B are afflicted. Country B objects to this cross-border diffusion of pollution, but the cost of doing anything about it is unaffordable to country A. Country A argues that the atmosphere, like the high seas, is free for all to use and cannot be compartmentalized by boundary planes. Still, country B may press for damages. The stage is then set for still another boundary-related dispute.

Such a scenario is not imaginary. Canada has complained vigorously to the United States about acid rain caused by pollutants emanating from the U.S. manufacturing belt; Scandinavia's forests and lakes have been affected by acid rain resulting from industrial pollution in Britain and mainland Western Europe.

To the many problems of political boundaries at the surface, therefore, we must add those above and below the ground. Hence we have even more reason to regret our tendency to be as territorial as we are.

HOW BOUNDARIES ARE MADE

If you have ever looked at a document entitled Property Survey and Description, you have a good notion of the complexity of boundary making. Imagine: a small property, a piece of land half an acre or

ACID RAIN: INTERNATIONAL BOUNDARY ISSUE

One of the most discussed environmental perils of recent years is acid rain, caused by the considerable quantities of sulfur dioxide and nitrogen oxides released into the atmosphere as fossil fuels (coal, oil, natural gas) are burned. These pollutants combine with water vapor contained in the air to form dilute solutions of sulfuric and nitric acids, which are subsequently washed out of the atmosphere by rain or other types of precipitation such as fog and snow.

Although acid rain usually consists of relatively mild acids, they are sufficiently caustic to do great harm to certain natural ecosystems. There is much evidence that this deposition of acid is causing lakes and streams to acidify (resulting in fish kills), forests to become stunted in their growth, and acid-sensitive crops to die in affected areas.

In cities, the corrosion of buildings and monuments is both exacerbated and accelerated. To some extent, acid rain has always been present in certain humid environments, originating from such natural events as volcanic eruptions, forest fires, and even the bacterial decomposition of dead organisms. However, during the past century, as the global Industrial Revolution has spread ever more widely, the destructive capabilities of natural acid rain have been greatly enhanced by human activities.

The geography of acid rain is most closely associated with patterns of industrial agglomeration and middle- to long-distance wind flows. The highest densities of coal and oil burning are associated with major concentrations of heavy manufacturing, such as those in Britain and Western Europe. As these industrial areas began to experience increasingly severe air pollution problems in the second half of the twentieth century, many nations, including the United States in 1970, enacted environmental legislation to establish minimal clean air standards. For industry, the easiest solution has often been the construction of very tall smokestacks (1,000 feet [300 meters] or higher is now quite common) that disperse pollutants from source areas via higher-level winds. These long-distance winds have been effective as transporters—with the result that more distant areas have become dumping grounds for sulfur- and nitrogen-oxide wastes. Regional wind flows all too frequently steer these acid rain ingredients to wilderness areas, where livelihoods depend heavily on tourism, agriculture, fishing, and forestry.

The spatial distribution of acid rain within Europe offers a classical demonstration of this. High sulfur emission sources are located in the leading manufacturing complexes of England, France, Belgium, Germany, Poland, and the Czech Republic. Prevailing winds from these areas converge northward toward Scandinavia, with a particularly severe acid rain crisis occurring in southern Norway. Lake acidity there is already in the moderately caustic range, and most fish species, the phytoplankton they feed on, and numerous aquatic plants have been obliterated.

Scandinavia is not alone in experiencing these environmental problems: at least 25 percent of Europe's forests have become affected. In Central Europe, forest ecosystems are now so badly damaged by acid rain that vast woodland areas are diseased and dying. More than one-third of Austria's forests are threatened; in Switzerland, the proportion nears 50 percent. The worst devastation, however, has occurred in Germany, where over two-thirds of the woodlands are afflicted and entire forests, such as the once lovely Black Forest in the country's southwestern corner, are all but decimated. Government agencies have belatedly recognized this ecological disaster, but their proposed remedies may well prove to be a case of too little too late.

less, may have to be described in several detailed paragraphs—not once, but each time it is sold. I remember buying a house in Georgetown (Washington, D.C.) some years ago. It was a rowhouse with a tiny garden in the back. Two surveyors spent most of a day working up the survey and description, which, weeks later, was part of the closing documents required to complete the purchase. Then, less than a year later, I refinanced the house. Believe it or not, another survey was done. The house hadn't moved, no construction had occurred, everything was as it had been. But the property was mapped once more.

If a tiny property such as mine could require such meticulous attention, imagine what is needed to make legal the surface boundary between two countries! And that's exactly what those surveyors were doing: they prepared the legal description of the bounds of my property. For an international boundary to have validity, it needs definition in legal terms. This means that it must, in the first instance, be described in terms of the physical and cultural features that mark or affect it. Before a mapmaker can draw a single line, before governments can drive a single post into the ground, the boundary's legal definition must be in place.

One might conclude that this has surely been done over most of the world. How many miles of boundaries could still be in contention? The answer is: many thousands.

Here's an example. If you look at virtually any map of the Arabian Peninsula (page 144), you will see that the political boundary between that region's two most populous states, Saudi Arabia and Yemen, is marked as "undefined," "indefinite," or "disputed." It is likely to be shown as a dashed line; compare several maps, and the boundary is likely to be in a different position on each one. In short, this is one of those boundaries that has never been defined. Worse, the two countries it separates have an unfriendly relationship. They were on opposite sides during the Gulf War, and while one is an autocratic kingdom, the other is a fledgling democracy. Still worse,

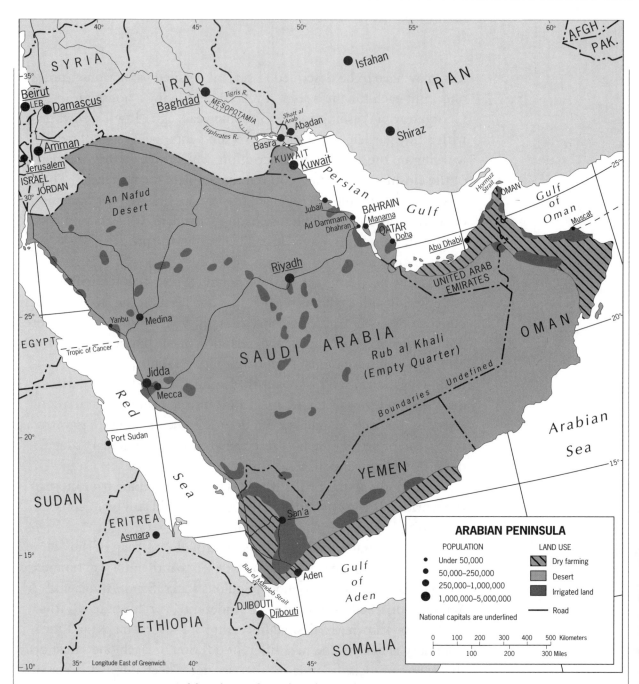

ARABIAN PENINSULA

POPULATION

- • Under 50,000
- • 50,000–250,000
- ● 250,000–1,000,000
- ● 1,000,000–5,000,000

LAND USE

- ▨ Dry farming
- ░ Desert
- ▨ Irrigated land
- — Road

National capitals are underlined

| 0 | 100 | 200 | 300 | 400 | 500 | Kilometers |

| 0 | 100 | 200 | 300 | Miles |

oil has been found right in the area where a boundary should long
ago have been defined. The prospect: strife. When the Gulf War
broke out, Saudi Arabia summarily expelled more than a million

Yemenis, forcing them to cross the empty strip where the boundary should lie. And when Yemen awarded exploration rights to oil companies working in the border area, Saudi Arabia threatened action and forced the companies to cease their activities. The absence of an agreed, defined boundary here is a serious threat to future stability.

Let's look at the bright side: the majority of international boundaries *are* defined, though some of them inadequately, and adjoining countries agree on their legal descriptions. What happens next? Once satisfactorily defined, a boundary is *delimited*, that is, it's drawn on official maps. Interpreting the legal definition can be quite a challenge for cartographers, and it is not uncommon for boundary maps to be drawn and redrawn until the parties agree that the line represents what the treaty makers had in mind. You know how lawyers write: I still don't understand everything that's in my property description.

Some governments don't wait for proper delimitation to take its course: they publish maps that delimit their version of a boundary agreement, not the joint version. That is, in two words, cartographic aggression, and it's more common than one might assume. China routinely publishes maps showing its boundary with India to lie well within Indian territory. Libya's official maps include a sizeable part of neighboring Chad. In Middle and South America, where many boundary disputes simmer, these cartographic wars have been going on for centuries.

The third and final stage of boundary making involves marking the boundary on the ground, a process called *demarcation*. Such demarcation can have different aims. Friendly neighbors may do little more than erect a concrete pillar at varying distances along their common border. The rule is that when you stand at one such pillar, you should be able to see the next one in each direction. Less friendly neighbors often attempt to create physical barriers along their common boundaries in order to control and inhibit movement. The Berlin Wall and the armistice line across the

TREATY LANGUAGE

If you want to get a backstage look at the way boundaries have been established, look up a book called *The Map of Africa by Treaty*, a two-volume work by E. Hertslet. It's full of incredible stories of colonial wrangling and shenanigans leading to the now familiar African political framework. Treaty language of necessity can be quite folksy. A boundary may be defined to run

...from the fencepost on the southeast corner of the property owned by William Morris Davis approximately due eastward along the crest of the ridge locally known as Bowman's Rise 5783 feet (1762 meters) to the center of the watergap occupied by the Semple River. The boundary follows the centerline of the Semple River southward (downstream) approximately 6.13 miles (9.9 kilometers) to its confluence with

the Ratzel Creek at which point the boundary becomes the west bank of the river. Precisely 695 feet (211.836 meters) from this point the boundary is marked by the center of the large Baobab tree from where it turns 135 degrees in a straight line for 3.47 miles (5.58 kilometers)...

When cartographers later translate this language into a functional map, they need details the map—even a large-scale map —may not show. The fencepost may be gone. The Baobab tree may have died. Erosion may have shifted the streams, invalidating the measures. In such instances, the cartographers and surveyors, perhaps not the original definers, may have to go to the field to determine the intent of the treaty makers. Even then, it may be difficult to satisfy the affected neighbors.

Korean Peninsula are a recent and a current example of this. Anyone who saw the concrete, barbed wire-reinforced Berlin Wall must have marveled at the fact that some people actually managed to get across it alive. The Korean boundary continues to cut a nation in half, and families asunder, an almost hermetic seal between two neighboring societies.

Demarcating boundaries for particular purposes is tempting as national policy. The United States is not immune: relations with friendly neighbor Mexico are generally good, but that has not stopped Washington from reinforcing its boundary in order to stem the tide of illegal immigration. Control posts on roadways, fences, patrols, and other efforts to strengthen the Mexican border all form part of our demarcation effort. Meanwhile, it is not uncommon for people to commute across the border with our northern neighbor, Canada, with little surveillance or inconvenience.

Every mile or kilometer of boundary, therefore, has a history and

BOUNDARY AND FRONTIER

News reports often use the terms boundary and frontier interchangeably. Chalk it up to geographic illiteracy: a boundary is a line on the ground that marks the vertical plane separating two countries, while a frontier is a surface area of undefined extent that separates two countries. Centuries ago, before many of the world's international boundaries were drawn and mapped,

frontiers—no man's lands—were common. Dense forests, swamps, deserts, and mountains served as barriers between nations, and boundaries were not urgently required. But today, most such frontiers have been carved up among neighbors. A few true frontiers remain, for example, that between Saudi Arabia and Yemen. But most frontiers aren't geography—they're history.

a purpose. The history, as is so often the case, is a matter of geography. The purpose is likely to reflect whether the adjacent countries are good and friendly neighbors—or not.

BOUNDARIES ON THE MAP

Even the most cursory look at a map of the world's countries raises the obvious question: Why are some boundaries simply straight lines, while others follow natural landscape features, and still others seem to wiggle about for no apparent reason?

The answer lies in the boundaries' genesis. Centuries ago, linear features in the natural landscape—rivers and mountain ranges —became trespass lines and acquired the status of political boundaries when modern nation-states arose. Still today, the Pyrenees form most of the boundary between Spain and France, the Danube River separates Romania from Bulgaria, and Andean crestlines partition southern South America between Chile and Argentina. Look at a large-scale map of such boundary zones, and the bends in the borders can be seen to coincide with nature's wiggles.

But if you took a National Geographic world map and traced with a magic marker all political boundaries that correspond with physical features, you'd still be left with thousands of miles of boundaries *not* natural in origin. What caused the wiggles in those borders? In fact, we can witness the process in progress during the present decade, in the former Yugoslavia, in Transcaucasia, in Cyprus, in Sri Lanka, and in many other places. Our world is a mosaic of ethnic

CYPRUS: GREEN LINE BOUNDARIES

In the northeastern corner of the Mediterranean Sea, much farther from Greece than from Turkey or Syria, lies Cyprus (population 740,000). Since ancient times, this island has been predominantly Greek in its population, but in 1571 it was conquered by the Turks, under whose control it remained until 1878. Following the decay of the Ottoman state, the British took control of the island. By the time Britain was prepared to offer independence to much of its empire in the years following the Second World War, it had a problem in Cyprus. The Greek majority among the population of 600,000 preferred enosis—union with Greece. It is not difficult to understand that the 20 percent of the Cypriot population that was Turkish wanted no such union.

By 1955, the dispute had reached the stage of violence, because differences between the Greeks and Turks are deep, bitter, and intense. It had become impossible to find a solution because the residents of this island country think of themselves as Greeks or Turks first rather than Cypriots. Nonetheless, in 1960 Cyprus was granted independence under a complicated constitution designed to permit majority rule, but with a guarantee of minority rights.

The fragile order finally broke down in 1974. As a civil war engulfed the island, Turkish armed forces intervened (about 30,000 soldiers were still present in the early 1990s), and a major redistribution of population occurred. The northern 40 percent of Cyprus became the stronghold of more than 100,000 Turkish Cypriots; only a few thousand Greeks remained there. The rest of the island—south of the United Nations-patrolled Green Line that was demarcated across the country to divide the two ethnic communities—became the domain of the Greek majority, now including nearly 200,000 in refugee camps. The fundamental differences between Greek and Turkish Cypriots are still unresolved, hatreds have deepened, and a form of de facto apartheid now exists on the island.

In 1983, again bringing the stalemated conflict to the attention of the world, the Turkish community, today comprising one-fifth of the population, seceded by declaring itself the new independent Turkish Republic of Northern Cyprus. The situation is especially difficult because of the many factions involved. The Turkish Cypriots, many of whom have expressed a desire for reintegration, are discrete from the Turkish nation and express a strong attachment to their island homeland. The Greek Cypriot majority is ideologically divided, and the crisis of 1974 began, not as a Greek-Turkish conflict, but as a result of a coup by Greek Cypriot soldiers against the Greek-dominated government of the island. Internal factionalism and external involvement have made Cyprus a pawn of foreign powers—and a casualty of history.

and cultural groups, and human history is the story of struggles for territory. Those bends and turns in boundaries from Western Europe to East Asia result from repeated, if not endless, clashes over turf. Even the smallest nations have battled over the tiniest fragments of land, resulting in boundaries that seem, in places, (may I use a subjective word?) ridiculous. A large-scale map of the border area

BORDERLINE STORIES

Every geographer probably has a story to tell about a border crossing or a boundary experience. My funniest moment happened in July 1969 at a place called Namanga on the Kenya-Tanzania border. I was in the field with a graduate student, later a Professor of Geography at the University of Miami, D. L. Capone. In my experience, African border officials are not especially gracious, and when we walked into the Tanzanian office I had my passport scrutinized, stamped, and returned without pleasantries. Capone was next.

As I stood by the door waiting for him, he laid his passport on the table. The immigration officer, in full uniform, looked down, then up at Capone, then down again. Next he rose, kicking his chair backward, and saluting elaborately. The look on Capone's face was worth the entire research grant. Then our official spoke, "Welcome to our country, Mr. Capone. On behalf of all Tanzanians, I hope you enjoy your visit." With that, he stamped Capone's passport, walked with him to the door, and waved us—no, him—farewell.

Only several days later, when we were able to stop at a hotel for a shower and a cooked meal, did we realize what lay behind the officer's remarkable gesture. On Tanzania's national television flickered a program being shown, it turned out, in nightly installments: *The Untouchables*. There must have been some confusion among Tanzanians as to who the hero was.

between Belgium and the Netherlands reveals tiny, village-size pockets of one country encircled by the other. It all harks back to the days when religion and language and other cultural traditions provoked border conflicts that surged back and forth across what was then an unstable frontier.

Abroad as well as at home, European nations as colonial powers took each other on and fought over territory—Spanish against Portuguese, British against French. In the colonized world, boundaries began to look as they did in Europe, each a testimonial to territorial competition.

So we can identify natural political boundaries and, for want of a better word, cultural ones—and we can see new cultural boundaries in the making, as in Bosnia, Cyprus, Sri Lanka, Kashmir, Transcaucasia, and elsewhere. When the cultural conflicts there reach a stalemate, the task of defining, delimiting, and demarcating the new boundaries will begin. Before long, atlases and wall maps will carry the imprints of these struggles. But this leaves still another kind of boundary, the straight line or geometric boundary. Americans ought to know about these: one of the world's longest geometric borders separates the

United States and Canada west of the Great Lakes, and part of our border with Mexico is geometric, too.

Geometric boundaries are boundaries of convenience. They do not require complicated legal definition: they only need coordinates. Since the late 1700s, the United States had taken the 49th parallel north latitude as the boundary between the French possessions centered on Louisiana, that is, the whole Mississippi Basin, and the British domain to the north. When, in 1803, France sold Louisiana to the United States, maps began to show the same parallel, west of the west coast of Lake of the Woods, as the U.S.-British border. At first, the British would have none of it, but in 1818 they acquiesced, arguing only over the western (Pacific) end of the boundary (they wanted it bent to reach the coast at the mouth of the Columbia River). They gave up that notion in 1846, and the boundary treaty has stipulated the 49th parallel, all the way to the Strait of Georgia, ever since.

In those days, nearly two centuries ago, few whites had penetrated this border region, and of course Native American domains or settlements were not taken into consideration. There seemed to be little reason to elaborately define, delimit, and demarcate a boundary that could be construed in one sentence. Ever since, a boundary that was there long before the patterns of modern settlement emerged has been in effect.

Later during the nineteenth century, Africa's boundary framework took shape, and in much of the northern regions of the continent, the colonial powers drew geometric boundaries to allocate their respective holdings. Vast stretches of Sahara were compartmentalized this way; why should empty wastes traversed only by nomads be treated any other way? Or so the colonial powers reasoned. And so Egypt became, literally, the northeast corner of Africa and Mauritania a nearly rectangular subdivision of French West Africa. Lands in between—Mali, Niger, Chad—were laid out in the same manner. The map of Africa is full of geometric boundaries: between Kenya and Tanzania, Angola and Namibia, Ethiopia and Somalia.

Times change, however, and empty spaces have a way of filling up. Africa and other areas of the world endowed with geometric borders have paid a price for this colonial legacy. In the northern Sahara, geometric boundaries happened to lie where oil and gas were later found, and since geometric boundaries were mostly undemarcated, conflicts over ownership soon arose. Elsewhere, peoples of common cultural heritage found themselves on opposite sides of geometric borders, and when governments got around to demarcating these borders, they sometimes cut villages in half. Maybe all that troublesome defining and delimiting is worth the trouble after all.

BOUNDARIES ARE HERE TO STAY

Here is an old lesson in geography: no matter how unreasonably placed, boundaries, once established, tend to become entrenched and are very difficult to relocate or eliminate. Of the tens of thousands of miles of world boundaries, very few have actually been erased. Exceptions include East-West Germany, North-South Vietnam, British-Italian Somaliland, Morocco-Western Sahara, East-West Timor. When African countries threw off the yoke of colonialism, one of their first postindependence agreements was to respect the boundaries with which they were endowed, for good or ill. Even in comparatively wealthy, unifying Western Europe, the European Union program is bedeviled by member countries' reluctance to let go of their boundaries and what they have meant to the nations. No, it's much easier to make more boundaries than to remove them. The number of member states of the United Nations is edging closer to 200, and our already limited earthly living space continues to be fragmented further. We may be living in a technological, postindustrial age, but our territorial imperative has hardly waned.

CHAPTER 8

PEOPLE, PEOPLE EVERYWHERE

Recently I found myself trapped in one of Northern Virginia's massive traffic jams. Cars stood on both sides of mine, in front, and behind. Lines in all lanes stretched as far as the eye could see. It was a cool day, and most drivers had their windows down. A fellow in a Mercedes next to me leaned over.

"Must be the population explosion! It's hit us, and we'll all expire!" he exclaimed. He must have got a noseful of the exhaust fumes now swirling around, because he raised his windows and turned on his air conditioning. So did almost everyone else: the lineup of idling engines created a soupy atmosphere on the nearly windless afternoon.

My neighbor was joking, but he had a point. If there is anyplace in America where you can experience for yourself the impact of rapid population growth, it's in the suburbs. Where I sat, outside Washington, D.C., forests of newly built townhouses stretched in all directions, having gobbled up meadows and farmlands. Ugly pylons and wires carried electricity to the new settlements. Construction crews were at work on roads, sidewalks, drains. And the traffic moved at a snail's pace, if at all.

"Is it like this every day?" I asked a gas station attendant up the road.

"It's usually worse," he said. "It's pretty quiet today, people aren't even honking their horns much. I'm from the Los Angeles area, and to me, this is nothing."

Among the nearly 200 countries in the world, the United States's population is not growing nearly as fast as many others. And it would be growing even more slowly were it not for the large number of immigrants, legal and illegal, that swell U.S. numbers. Still, the pressure of population is felt even here, in spacious America. Suburbanization has created vast rings of urban sprawl around original cities, and in those urban peripheries, population growth and influx have overtaken infrastructures. In parts of suburban America, the very irritations of crowding that impelled many to move out of the city have caught up with the movers.

THE GLOBAL SPIRAL

Sitting in a traffic jam is a small price to pay for population growth. Elsewhere in the world, people are starving or suffering from the permanent ravages of malnutrition. By the hundreds of millions, these people pay the price of being in the wrong place at the wrong time. Because the truth is that all people on Earth today *could* be fed—not well, but adequately for survival—if there were ways to distribute available supplies to all. Unfortunately, that is not happening. In Somalia, competing warlords were perfectly prepared to starve their competitors to death. President George Bush, moved as were many Americans by television pictures of the dying, sent armed forces and food to the rescue. Not seen by television viewers was the far more deadly situation in Sudan, where no such help was or could be given. From Angola to Liberia to Afghanistan, hunger resulting from dislocation is a constant companion.

As recently as the 1970s, the specter of inevitable and permanent food shortages still loomed, and it was impossible to predict that the Earth's total food supply could actually catch up with the ever-increasing demand. But technological developments in farming raised production to unprecedented levels, and by the 1980s the fear of global hunger had receded. Today, the food problem is more geographic than anything else. Whether you'll go through life adequately nourished is a matter of location, of where you happened to be born. The powerless and influenceless peoples of the African and Asian interiors are the hungriest. But

A STABILIZING WORLD POPULATION?

Numerous national and international agencies monitor trends affecting world population growth, including the United Nations, the UN Food and Agricultural Organization (FAO), the Population Reference Bureau, the World Bank, and the U.S. Census Bureau. Each of these agencies publishes reports on its findings, and from these an occasional ray of hope arises.

A recent report by the World Bank suggests that world population will continue its rapid growth for some time but that its rate of increase will begin to slow during the twenty-first century. By the year 2000, according to this report, India will have 958 million people, Mexico 126 million, Brazil 205 million, Indonesia 198 million, Pakistan 135 million, Nigeria 154 million, and Bangladesh 146 million.

However, the report predicts that all these countries will reach a stationary population level, just as some Western European countries have. The United States will stabilize at 276 million people in 2035. Other projections include Brazil (353 million in 2070), Mexico (254 million in 2075), China (1.4 billion in 2090), and India, destined to become the world's most populous country (1.6 billion in 2150).

Such long-range predictions have uncertain foundations. Conditions in the populous underdeveloped countries today are not the same as those that prevailed in Western Europe during its nineteenth-century population explosion, so the aftermath may well be quite different. However, some researchers see sufficient evidence of a long-range decline in population growth rates to believe that they can predict it accurately.

India, for example, is now able to feed itself under normal circumstances, an achievement few foresaw just a generation ago.

This does not mean that the future is risk-free. Major environmental changes over the short term could reduce the output from world breadbaskets such as the U.S. Midwest or the North China Plain. Unanticipated outbreaks of plant diseases could set back the ongoing agricultural revolution. And the inexorable march of population growth could once again outstrip the Earth's productive capacity. About 70 percent of our planet's surface is covered by water or ice, and of the remaining 30 percent, two-thirds is arid, frigid, mountainous, or otherwise inhospitable to human habitation. That, however, has not stopped our species from multiplying like a bacterial organism in an optimal environment.

In the mid-1990s, about 180 million babies are born every year, and about 85 million people die, many of them at birth. This means that we are adding about 95 million to our number every year: nearly 8 million per month, 270,000 per day, 11,000 per hour! The Earth's population is now rushing toward 6 billion; if some killer epidemic or natural disaster or nuclear war does not slow us down, there will be 10 billion—10,000 million—people on this planet around the year 2030. Many of my colleagues in the field of population geography see this population explosion as our greatest challenge; some would divert much of the money now going toward environmental protection to population control programs.

But not everyone agrees. There are those who argue that the world must pass through this period of explosive population growth, with all its destructive consequences, in order to reach a point where it will fizzle out—where the world's population will stabilize. The World Bank, in its annual *World Development Report*, even publishes a table giving the date when individual countries' populations are projected to cease growing. For the moment, though, world population as a whole continues to erupt, producing streams of numbers unimagined just a century ago.

POPULATION AND POLITICS

In June 1992, a United Nations Conference on Environment and Development was held in Brasilia, the capital of Brazil. When I saw the agenda, I was surprised at the lack of attention given to the population issue. When I saw the roster of participants (delegations consisted of political leaders, scholars, and hangers-on) I was less surprised: very few were geographers. The population question was not a salient topic, but it should have been. There's little point in making commitments to protect what remains of global natural environments without taking the demographics into account.

It will be remembered that President Bush came under heavy criticism at home and abroad for not providing sufficiently strong leadership on major environmental issues discussed at the conference, and the American delegation generally had an uneasy time. But there was plenty of censure to go around. The Roman Catholic Church is strong in most of Middle and South America, and the church remains a major force in opposition to birth control as a means of family planning. Judging by the statistics, Catholics in Europe pay little attention to such strictures: Italy is one of the world's slowest-growing countries. But underdeveloped Middle and South America are something else. During a visit to Colombia in 1986, Pope John Paul II reiterated his opposition to artificial birth control and in effect exhorted Colombians, and many millions of listeners throughout the realm, to have as many children as they wished. Brazil, the host country for the conference six years later, also remains a strongly traditional Catholic society. This was not a good place to promote practices that might have a more salutary effect on future global environments than all the pollution-reduction programs put together.

This does not mean that U.S. policy on population has been focused or steady. Successive administrations have alternately helped and hindered family-planning programs here and elsewhere.

And yet, what is happening to population growth worldwide should be of direct concern to all Americans. Underdeveloped countries coping with high growth rates see their hopes of improved living standards dashed and their dependence on foreign help raised. Overpopulation leads to economic conditions that generate desperate emigrations—and the migrants often cross American borders. Mushrooming populations stress the capacities of governments to exercise control—and American relief efforts plus, as in Somalia, policing campaigns cost lives and money. Lately, the American public has begun to worry about these issues, and the immigration question is moving higher up on the national agenda. The overwhelming approval of Proposition 187 in California during the November 1994 elections reflects a rising anti-immigrant feeling in a nation forged of immigrants.

POPULATION AND THE ENVIRONMENT

Let's focus for a moment on the impact of humanity's growing numbers on the natural environment of Earth. Biologists estimate that there may be as many as 25 million types of organisms on Earth, perhaps even more; most have yet to be identified, classified, or studied. *Homo sapiens* is only one of these, and in 10 millennia our species has developed a complex culture that is transmitted from generation to generation by learning and is to some degree encoded in our genes. We are not unique in possessing a culture: gorillas, chimpanzees, and dolphins have cultures, too. But ours is the only species with a vast and complex array of artifacts, technologies, laws, and belief systems.

No species, not even the powerful dinosaurs of epochs past, ever affected earthly environments as strongly as humans do today. The dinosaurs, and many other species, were extinguished by what may have been an asteroid impact combined with huge volcanic eruptions. Some biogeographers see an analogy and suggest that the next great extinction may be in the offing, caused not by asteroids but by humans, whose numbers and demands are destroying millions of species—and with them the inherited biodiversity of this planet.

This destructiveness is not just a matter of modern technology and its capacity to do unprecedented damage, whether by wartime forest defoliation, peacetime oil spills, or other means. Human destructiveness manifested itself very early, when fires were set to kill whole herds of reindeer and bison, and entire species of large mammals were hunted to extinction by surprisingly few humans. The Maori, who arrived in New Zealand not much more than 1,000 years ago, inflicted massive destruction on the native species of animals and plants in their island habitat, long before modern technology developed more efficient means of extinction. Elsewhere in the Pacific realm, Polynesians reduced the forest cover to brush and, with their penchant for wearing feather robes, had exterminated more than 80 percent of the regional bird species by the time the first Europeans arrived. The Europeans proceeded to ravage species ranging from snakes to leopards. Traditional as well as modern societies have had devastating impacts on their ecologies, and on the ecologies of areas into which they migrated.

Is wanton destruction of life a part of human nature, whatever a society's cultural roots? The question is as sensitive as questions about racism and sexism. Still, regional differences in attitude and behavior can be discerned. African traditional societies hunted for food or for ceremonial reasons, but not for entertainment or amusement. The notion of killing for fun and fashion was introduced by Europeans. Hindu society and religious culture in India are more protective of the natural world than many others. The extermination and near-extermination of many Indian species of animals took place during (Muslim) Moghul and European colonial times.

Malevolent destruction of the environment continues in various—indeed many—forms today, ranging from the deliberate spilling of oil and setting of oil fires by Iraqis during the 1991 conflict over Kuwait, to the mercury poisoning of Amazonian streams by Brazilian gold miners. For the first time in human history, however, the combined impacts of humanity's destructive and exploitative actions is threatening the entire Earth's biodiversity. Most of that biodiversity has been concentrated in, and protected by, the great equatorial and tropical rainforests of South America, Africa, and Southern Asia. Now the onslaught on this last biogeographical frontier is under way, and for the future of the planet, the consequences may be catastrophic.

EXPLODING POPULATIONS

The explosion of global population during the past two centuries, and especially during the twentieth century, is common knowledge now. About 200 years ago, shortly after the onset of the Industrial Revolution, the Earth's total population was only about 900 million, which is the present-day population of India alone. In 1820 the milestone of 1 billion was reached, and this number took more than a century, 110 years to be exact, to double. Then the upward spiral really took off. Only 45 years later (1975), 2 billion doubled to 4 billion, and at present growth rates, this number will again double, to 8 billion, around the year 2010. We're halfway there: the 6 billion mark is being passed in the mid-1990s.

There are times when scary statistics are suspect, but not in this case. For thousands of years of human evolution and civilization's development, our numbers remained small, to be measured in tens, then in hundreds of millions. Even as recently as 1650, the world's human population was about 500 million—as many as are being added now every five years or so.

Why is this happening? A combination of factors caused what is, in truth, explosive population growth. A population's natural increase over a given period is measured, simply, by subtracting deaths from births. When the number of deaths per 1,000 people in the population is close to the number of births, that population will grow very slowly. But when the number of deaths declines, the gap widens and population growth increases. Taking the world as a whole, death rates began to decline during the eighteenth century, while birth rates remained high. That created a widening gap, and while birth rates also started to decline in some parts of the world, this decline lagged far behind the death rate. The lag is still in effect.

Death rates declined because of major progress in hygiene and medicine associated with the Industrial Revolution and because of their export from the source (mainly Europe) to the rest of the world. Just two inventions, effective soap and the toilet, contributed enormously to this; ever-better medicines had great impact as well. Refrigeration, water purification, and other advances also helped lower the death rate. Such progress ensured the survival of countless babies who would otherwise have died at or soon after birth. The death rate of a population reflects not only those who die after a lifetime, but also those who die at birth or in infancy. Today, one of the most telling statistics reflecting a country's overall condition is its infant mortality rate. In the United States, infant mortality is slightly over nine per thousand births annually—by no means the world's lowest. In 10 African countries it still exceeds 125 deaths per thousand births.

POPULATION GEOGRAPHY

This brings us to the geography of demography: the regional differences in population growth. A list of population growth rates in individual countries (see Appendices: "Area and Demographic Data for the World's States") reveals the huge range from one part of the world to another. As noted earlier, European countries where the Catholic religion is dominant are not the fastest-growing by any means. Italy, Spain, and Portugal are growing at only 0.2 percent or less annually, rates that rank among the world's lowest. In fact, some European countries have shown negative population growth in some years. Not counting their annual quota of immigrants, such countries as Denmark, Germany (West Germany previously), and Belgium have had a *decrease* in population. The very influx of immigrants into Western European countries has been a corollary of the region's nearly static population: North Africans and other foreigners came by the millions to do the jobs for which locals are not available. Now, with European economies in decline, those foreigners are the targets of extremists. Germany's large Turkish population, in particular, has suffered from neo-Nazi anti-immigrant violence.

As a geographic realm, Europe today is experiencing the lowest rate of natural population increase in the world. Virtually all of Europe has a growth rate well below 1 percent annually; the United States, Canada, and Australia (and, if we can believe available statistics, Russia and its neighbors) are also in this category. By contrast, the highest growth rates today are recorded in Africa and Southwest Asia, where many countries report increases in excess of 3 percent per year. Muslim countries, in particular, are growing extremely rapidly. Pakistan, for example, is growing much faster than India, and populous Iran is growing faster than Pakistan. But even higher rates of natural increase are reported from eastern Africa. There, during the 1980s, Kenya actually reached a 4 per cent growth rate, which meant that its population was doubling every 17 years. In the mid-1990s, Kenya's growth rate, while still among the highest in the world, was down to 3.8 percent and declining, still slightly higher than neighboring Tanzania and, beyond, Zambia. Uganda, Rwanda, Burundi, and Zaire, also growing at very high rates, are likewise showing signs of decline.

DOUBLING TIME

If you want a vivid picture of the meaning of percentage growth rates, there's no better way than to convert these to *doubling time*, the number of years it would take for a population to reach twice its current size. The Earth's population as a whole is currently growing at about 1.8 percent per year, which gives a doubling time of just under 40 years. So if the growth rate doesn't drop, a baby born today will see the Earth's human numbers double twice in his or her 80-year lifetime: from 6 to 12 to 24 billion.

Some individual countries' populations are growing even faster. In East Africa, where Kenya and its neighbors grow at rates ranging from 3.5 to 3.8 percent annually, the doubling times are a mere 20 years or so. The United States, growing at about 0.7 percent annually (not counting immigration) is now doubling its population in just under 100 years.

POVERTY AND AIDS

Two questions arise. First, East African states are very poor. How can such high growth rates, so many children, be accommodated? And second, this region is severely afflicted by the AIDS pandemic. Why are growth rates still so high when so many people are sick with AIDS?

Let's look at the question of poverty and procreation first. Not just in East Africa, but worldwide it is and has long been true that poverty and hunger do not significantly or permanently reduce rates of natural increase. Cultural traditions are strong, especially in the countryside (when a society urbanizes, its growth rate is likely to decline). Women in the villages of Kenya and elsewhere in tropical Africa had, on average, seven, eight, even nine children. It was a

AIDS AND AFRICA

AIDS—acquired immune deficiency syndrome—became a pandemic in the early 1990s. In the mid-1990s, its impact on Africa was devastating, afflicting millions and creating prospects of negative population growth in some areas.

Persons infected with HIV (human immunodeficiency virus) do not immediately or even soon display visible symptoms of AIDS. In the early stages, only a blood test will reveal infection, and then only by indicating that the body is mobilizing antibodies to fight HIV. People can carry the virus for years without being aware of it; during that period, they can unwittingly transmit it to others. Official reports of actual cases of AIDS thus lag far behind the reservoir of those infected. In the United States, for instance, the Centers for Disease Control and Prevention in Atlanta had recorded about 250,000 cases as of mid-1993, but estimates of the number infected approached 1.5 million.

These U.S. data, however, pale before those from Africa. Official statistics for African countries still give no indication of the magnitude of the AIDS epidemic there, for obvious reasons. The medical system, already overwhelmed by the prevailing maladies of tropical Africa, cannot cope with this new onslaught. Many of those ill with AIDS live in remote villages or in the vast shantytowns of Nairobi, Kinshasa, and other large cities, and they do not see a doctor. Yet staggering evidence of the impact of AIDS is everywhere, and surveys by the World Health Organization have begun to unveil the real magnitude of the disease in Subsaharan Africa: over 6 million HIV cases as of late 1992.

Most disconcerting are the trends. Because AIDS is a blood and semen-borne disease and is transmitted most efficiently through (often undetected) blood contact, sexual practices in a society can presage routes of spatial diffusion. Patronage of prostitutes along the East Africa highway linking Kampala, Uganda, and Kenya's Nairobi and Mombasa, for example, is known to be one such diffusion pathway. In their 1991 book, *The Geography of AIDS*, medical geographers Gary Shannon, Gerald Pyle, and Rashid Bashshur reported a series of tests on female prostitutes in Nairobi: in 1981, 4 percent tested positive for the virus; in 1984, 61 percent; and in 1986, 85 percent. In Blantyre, Malawi's largest city, pregnant women in 1984 showed an infection rate of 2 percent; in 1990, the rate was 22 percent. About one-third of the babies born to infected women are themselves infected with the AIDS virus.

Infection rates in East and Equatorial Africa's cities are high, ranging from 8 percent of adults in Zaire's capital of Kinshasa to 30 percent in Kigali, capital of Rwanda. These levels have been reached relatively recently, so the real devastation of the epidemic in Subsaharan Africa has not yet begun. The cities were the hardest hit, but the rural areas have not escaped the ravages of AIDS. Surveys in rural areas of Uganda and Zaire indicate that between 8 and 12 percent of the adult population in those places are infected. The present state of medical knowledge holds out no long-term hope for those infected. Thus before the end of the century, AIDS will alter the population growth rates of much of tropical Africa.

matter of family security and status: several would die, the father wanted sons, and a high number of births improved the odds. While such customs continued, the death rate declined. Infant and child mortality in Africa still are horribly high, but they're lower than they were. Combine the continued fertility and the declining mortality, and you get skyrocketing growth rates.

The prospect that many of the children born will die early or will be malnourished is not a consideration when family survival is at stake and when only half the babies will be boys: the cycle of reproduction must continue. So growth rates stay high, even in grinding poverty. The impact of AIDS in tropical Africa has yet to reach its peak, but when it does, over the next decade, it will indeed reduce growth rates severely. Some observers suggest that the declines in growth rates seen in Uganda, Rwanda, and Burundi, as well as Zaire, slight as they may be, can be attributed in substantial part to AIDS. It is predicted that parts of tropical Africa, where AIDS is a heterosexual, society-wide disease, will actually record a decline in population, partly offset (and thus statistically hidden) by continued growth in less-affected areas. Already, AIDS and related causes are pushing up the death rates here, and as the pandemic takes its growing toll, death rates will actually overtake birth rates in parts of several countries. Quantitative predictions vary widely, but all agree that the overall population growth of tropical Africa will be reduced by millions. Estimates range from 12 to more than 30 million over the next 15 to 20 years.

GROUNDS FOR OPTIMISM?

For a more hopeful picture, we must look elsewhere on the map. Over the past two decades, Middle and South America's countries have shown a significant decline in population growth rates, despite religious strictures and generally slow economic development. In one country, Uruguay, the rate fell below 1 percent per year for the first time in history, and neighboring Argentina (1.2 percent in 1993) may be next. While several small Central American countries still have high growth rates, giant Brazil is now well below 2.0 percent, and virtually everywhere in the realm, growth rates are dropping.

But the key countries are China and India, whose combined populations exceed one-third of the world's total. And there, too, the news is improving. China under Mao Zedong's communist regime encouraged large families, and China exploded demographically—but during the post-Mao period of communist "pragmatism," that policy was reversed. Chinese families were told to have only one child, a rule that was enforced by sometimes draconian methods. (It also contributed, according to many reports, to widespread female infanticide.) China's growth rate declined precipitously, from over 2.0 percent per year in the late 1960s to 1.1 percent in 1985. China's planners began to dream of a stable population of 1.2 billion by the year 2000. But then China's economic and social revolutions began to weaken the government's controls, and the growth rate inched upward again. Today, it stands at 1.3 percent per year—higher than the government would like, but still much lower than it was just two decades ago.

Unlike China, India, with over 900 million people, is a democracy, and while the government in New Delhi wanted to reduce growth rates, no national rules could simply be imposed to make it happen. Individual states of India tried to implement population-control measures, including tens of millions of sterilizations, but other means were more effective. And this brings up a point in connection with the world map: the figures it provides are necessarily national data, so that we can compare what is happening in individual countries. But, as India proves, population growth can vary widely within countries. In India, population-planning facilities and education were far more effective in some states than in others. As a result, the southern states' populations grow at rates well below the national average, while the north, and especially the northeast, grow much faster. Although data from the most recent census of India are preliminary, Indian colleagues tell me that these regional gaps are widening. They also report that the Muslim minority in India, at about 110 million the largest minority in any country in the world, is growing substantially faster than the population as a whole. Such regional and ethnic or cultural disparities in growth obviously have political implications. India's eastern and northern states are increasing their clout in national politics by virtue of their burgeoning populations; the rapid growth of the Muslim minority is adding fuel to the fire of the Hindu fundamentalist movement.

HOW MANY IS TOO MANY?

That's the key geographic question. Imagine for a moment a world in which the population is evenly distributed—no crowded cities, no empty deserts. In that world today, population density (the number of people per unit of area) would be just under 100 per square mile, giving every person more than six acres of land. That hardly seems to presage a standing room only world, except that population is not evenly distributed. The population densities of some countries are truly staggering: about 2,200 per square mile in Bangladesh, 1,500 in Taiwan, nearly 1,000 in the Netherlands. Some areas of the world are enormously crowded, others are virtually empty.

One part of the answer is, Where are people living? In cities or in the countryside? In Bangladesh, only a small minority of the people live in cities. Most of the rural land really *is* as crowded as the statistics suggest. But in the Netherlands, just a tiny minority lives on the land: almost everybody lives in towns and cities. There, the high density figure doesn't tell the whole story. So in addition to the density figure, we need to know how urbanized a society is. When we know that, we can tell better whether that many is too many.

Another part of the answer has to do with technology. What are people doing in those high-density countries? In Bangladesh, where most people live on the land, subsistence farming—farming for survival—is the norm. But in Taiwan, manufacturing is the rule and fac-

DISTRIBUTION OF WORLD POPULATION

The world's population is large and is currently growing at a rapid pace with no sign of slowing down. The 1995 total is approximately 5,800,000,000, with an annual rate of net increase of about 1.8 percent. Thus, by the turn of the century, more than 6.3 billion people will be alive, a gain of nearly 15 percent in only eight years. A more revealing measure of global population growth is its doubling time, the number of years required to increase its size to twice its present level. That figure today is 39 years. Therefore, 11 billion in 2031 will become 22 billion by 2070 and 44 billion by 2109—an eightfold increase over the next 117 years!

These predictions are especially alarming because the human world is already overcrowded and there is relatively little new living space to be opened up. The present ecumene (the permanently settled portion of the earth) has developed over many centuries and represents the totality of adjustments people have made to their environments. Thus, the most productive places have been occupied for generations, and their agriculture and other technologies have now been pushed to the limit to sustain large populations at close to the maximum carrying capacities of these fertile lands. And still the population keeps multiplying at a prodigious pace—a net increase of over 97 million a year, 8 million a month, 270,000 per day, or 11,250 per hour.

The distribution of world population is uneven. Fully 90 percent of humankind resides in the Northern Hemisphere. More than half of the people are concentrated on only 5 percent of the Earth's land, and nearly 90 percent live on less than one-fifth of the land. Because of the fertility of river valleys, plains, and deltas, lowlands contain the highest population densities, which decline markedly with increasing altitude. Well over half of humanity resides below the 600 feet (200 meter) level, and over three-quarters live below 1,650 feet (500 meters). Moreover, nearly 70 percent of the population lives within 320 miles (500 kilometers) of a seacoast.

Three major population clusters, which together contain about 65 percent of humanity, lie in East Asia, South Asia, and Europe; although considerably smaller, eastern North America is usually regarded as a fourth major cluster. We can also observe a number of minor population clusters, among them the Nile valley and delta of Egypt in northeastern Africa, several river delta areas of Southeast Asia, the Indonesian island of Java off the Southwest Asian mainland, parts of West Africa particularly Nigeria, and the southern Atlantic seaboard of Brazil in eastern South America.

This small number of concentrations encompasses over 80 percent of the world's population, and the remainder of the ecumene consists of much less densely settled areas that are less inviting in terms of topography, climate, and moisture availability.

tories, not farms, are the norm. When people can make and sell goods to the world, they free themselves from ties to the soil.

Take the case of Japan, a country about the size of Montana. Its archipelago is mountainous: there isn't much level land. Yet Japan has more than 125 million inhabitants, most of them clustered in several of the world's largest cities, including the largest of all, Tokyo;

the population of Montana is about 800,000. Japan went through a period of rapid population growth, but then the Japanese, with typical efficiency, went into the family-planning business and reduced the growth rate to near zero. Meanwhile, Japanese technology and productivity reached unprecedented, and in some ways unrivaled, levels. Raw materials poured in from around the world, because Japan had few itself. Oil was shipped in from the other side of the globe, because Japan has none. Farms lay vacant as workers streamed to the factories in and around the mushrooming cities. And from Japan to world markets streamed goods of every imaginable kind. From the population data, we might conclude that Japan is grossly overpopulated, but Japan's economic geography sets us straight. Japan makes all the money necessary to buy what it needs. No subsistence agriculture here.

Japan has just about as many people as Bangladesh, but the Japanese live, by some measures, 30 times as comfortably and well. Over a lifetime, a Japanese person will require and consume about 30 times what a person in Bangladesh consumes. Western Europeans consume still more, and Americans more yet. So when a Bangladeshi baby is born, prospects are that he or she will require far less of this Earth than an average American, European, or Japanese.

But what if the system were to collapse? What if the supply of raw materials to Japan were to fail, or if oil shipping were interrupted, or a major economic setback occurred, or a giant earthquake devastated its cities? Then Japan might well prove to be as vulnerable as Bangladesh to its frequent and disastrous cyclones, and Japanese recovery might be long in coming. Under any such calamity, Japan's huge population and its inability to support itself would create a world crisis. Technologically advanced societies may be relatively free of ties to the soil, but they are not free of risk.

So to the question How many is too many? there is no simple answer. When population experts first began to warn of impending doom (during the early nineteenth century), they could not have anticipated the Earth's capacity to provide for so many: more people live well and comfortably today than *existed* on the Earth a century ago. On the other hand, the scale of suffering also was unforeseen. For hundreds of millions of people, life is hungry, painful, harsh, and brief.

AMERICAN PATTERNS

Consider this: the United States, including Alaska, is almost exactly as large territorially as China, but its population is less than one-fourth of China's. And while China's rate of natural increase is 1.3 percent per year, that of the United States is little more than 0.7 percent. Population growth shouldn't be an issue in America.

But it is, and Americans are getting exercised about it. The reason is not natural increase, but another kind of population growth: immigration. Millions of people from Mexico, Central America, and the Caribbean, from across both the Pacific and the Atlantic are entering the United States, many as legal immigrants, many more illegally. Add these to the annual natural increase, and U.S. *total* population growth may approach 1 percent per year, lowering the doubling time to 70 years.

This influx of immigrants, plus the continuous shifting of the resident population, is changing the map of the United States. Earlier, we noted the nationwide movement from city to suburb; that shift is essentially without wider regional implications. But even while immigrants arrive on America's shores and cross its southern boundary, Americans are moving—from north to south, from Rustbelt to Sunbelt, from east to west. As a result, some States are growing faster, in fact much faster, than others, while a few actually have shrinking populations.

The general geographic pattern, with some exceptions, is this: States along the southeastern, southern, and western periphery of our country are gaining population, while States in the heartland and in the northeast are static, or nearly so, or losing.

In percentage terms, the fastest-growing State between the

Census years 1980 and 1990 (and this may surprise you) was Alaska, at nearly 37 percent. Arizona, with a nearly 35 percent increase, was next. Neither Alaska nor Arizona has a large resident population, so these high percentages do not represent large numbers.

In terms of total increase between 1980 and 1990, America's already most populous State, California, also grew the most—by more than 6 million people to almost 30 million. Florida was next, with well over 3 million additions, for a total of 13 million. Texas gained nearly 3 million residents and had 17 million in 1990.

Compare these numbers to those of our third-largest State, New York. In 1970, New York had 18.2 million residents. By 1980, that total had declined to 17.6 million. In 1990, it was just under 18 million—still well below the 1970 figure. In December 1994, Texas overtook New York and became the nation's second most populous State.

New York did gain a bit between 1980 and 1990, and immigration contributed to this. Other States continued to lose people: West Virginia lost 8 percent of its population, Iowa nearly 5 percent. In this world of rapid population growth, few entities the size of West Virginia or Iowa can report reductions.

There's nothing new about population shifts in America. The center of U.S. population has been moving steadily westward for 200 years, a movement that has developed a southward turn in recent decades. Today, that center lies about 65 miles (105 kilometers) southwest of St. Louis, Missouri; it crossed the Mississippi in 1980. Just one century ago, it lay in the southeast corner of Indiana.

The relative population gains and losses of individual States are immediately reflected by the growth or decline of their political representation in Washington, D.C. The size of a State's population determines the number of representatives it has in Congress, and if one State's delegation grows, another's must shrink: the total in the U.S. House of Representatives does not vary from 435. The big winner of 1990 was California, which gained seven representatives; Florida gained four and Texas gained three. New York lost three seats, and Pennsylvania, Ohio, Michigan, and Illinois each lost two. Other winners were Virginia, North Carolina, Georgia, Arizona, and Washington—every one a coastal or border State—while losers included West Virginia, Kentucky, Iowa, Kansas, and Montana, all interior states. The American population pattern continues to change, and with it the political landscape.

CROWDING'S CONSEQUENCES

Thomas Malthus, an English economist, published a warning in 1798 that population in Britain was growing faster than the means of subsistence. He predicted that population increase would be checked by hunger within 50 years, leading to the disintegration of the social order. For three decades after sounding the alarm, Malthus faced severe criticism from those who saw the future differently, but he gave as good as he got. The exchange is one of the most interesting debates ever printed. In the end, both Malthus and his critics were proven wrong: food production has not, as Malthus prophesied, increased in linear fashion. It has grown exponentially, although in spurts. And populations did stabilize, as they have in Western Europe.

Malthus and his colleagues argued over rates of increase and supplies of food, but they left untouched another consequence of explosive population growth: ethnic and cultural competition and strife.

By no means can all wars or conflicts be attributed to the crowding of our planet. When the European powers launched their campaigns of colonial annexation, the Earth—comparatively speaking—was a rather empty place. When Napoleon dreamed of a French-speaking world, there were fewer people on Earth than there are in India today. When Germany started the First World War, the world had only about one-quarter of the population it does today. Even Hitler, in the late 1930s, probably thought he'd have it easy: his geopoliticians told him that his global *reich* would have under 2.5 billion subjects.

Today, the numbers are very different. Clans have become tribes. Tribes have become nations. Nations have become powers. And growing numbers have occupied ever more space. Frontiers of separation have disappeared, and livable space is filled. Peoples who, with a little distance, might live in neighborly peace now glare at each other at close range. And all too often, the glare turns into warfare.

So it is or has recently been in Yugoslavia, Georgia, Sri Lanka, Cyprus, Ethiopia, Liberia, and Rwanda. While it may be technically true that our globe will be able to sustain even more billions in the future than it does today, what is the evidence that so many people, thrown so closely together on so small a planet, can overcome their ethnic, cultural, and traditional differences? Their territorialism? Their aggressiveness? While Western European and other national governments struggle to create multinational unions out of their fractious countries, ethnic or cultural groups within those countries, now large enough to see themselves as nations, seek greater autonomy, if not independence. And so the map of the world fractures into ever smaller pieces, the political fragments are admitted into the United Nations as full members, and hopes for a stable, unified world recede.

The population spiral is more than a matter of food supply or environmental degradation. Its impact on the political system may hold the real seeds of destruction.

CHAPTER 9

DIVIDING UP THE OCEANS

We humans are territorial creatures, and we have created a world honeycombed by boundaries ranging from Berlin Walls to backyard fences. And now we're in the process of dividing up the Earth's last frontier: the oceans. Today, many maps of individual countries display not just their land areas, but also their maritime domains. Maritime boundaries, in some parts of the world, are every bit as sensitive and as intensely disputed as boundaries on land. The parceling out of ocean space to coastal countries is a major development of our time, a crucial geographic transition.

OLD HABITS DIE HARD

We talk romantically of the high seas and of the open oceans as though these last frontiers are still free and unfettered, but in truth, the age of unrestrained ocean travel is over.

The end came quite quickly, in a matter of decades before and after the Second World War. But the notion of control over water (especially vital waters such as bays and straits) is centuries old. It all seems to have started in the North Sea and the western North Atlantic Ocean as long ago as the thirteenth and fourteenth centuries: the Dutch, the Norwegians, and the British, looking for the best fishing grounds, started to designate some of these marine areas as theirs alone. At first, their claims were rather general, but by the sixteenth century the Danes got into it—and announced that

Facing page: *The Truman Proclamation focused world attention on the potential of the continental shelves.*

a zone of water eight nautical miles wide off their coasts belonged to them alone. Since the Danes controlled not only Denmark but also the Norwegian coast and Iceland, that put a good-sized chunk of the North Atlantic off-limits to the other fishing nations of northwestern Europe. Soon the Danes extended their coastal zone to 24 nautical miles (a nautical mile is about 1.15 statute miles [1.85 kilometers]), and it looked as though the British, Dutch, and others would be squeezed out of the region altogether. The British retaliated by drawing imaginary baselines across their bays so that others could not fish near their coasts without violating British sovereignty.

But then the oceans' purpose changed. As the era of colonialism and mercantilism dawned, fishing became relatively less important and trade more so. Now the designated coastal zones might deter ships under foreign flags from entering trading ports, and the advantage of narrower coastal zones became apparent. Thus arose the notion of the territorial sea, a comparatively narrow strip of sovereign waters that would belong to any and all coastal states. In retrospect, the old competitors came close to general agreement on its width: the Danes reduced their claim to four nautical miles, the Dutch to three, and Spain and Portugal limited themselves to six. That hardly affected the world's open oceans: even six nautical miles is less than the thickness of the line showing the coast on a wall map of the world.

So things remained for centuries—in fact, until well into the twentieth century. Certainly, there were disputes over territorial waters, over occasional violations, over narrow straits, and inconvenient baselines. But as recently as 1951, about 90 percent of the world's merchant-shipping tonnage was registered in countries that restricted their territorial seas to the three-mile or four-mile limit. Yet the seeds of maritime territorial competition had been sown, and a scramble for the oceans lay ahead.

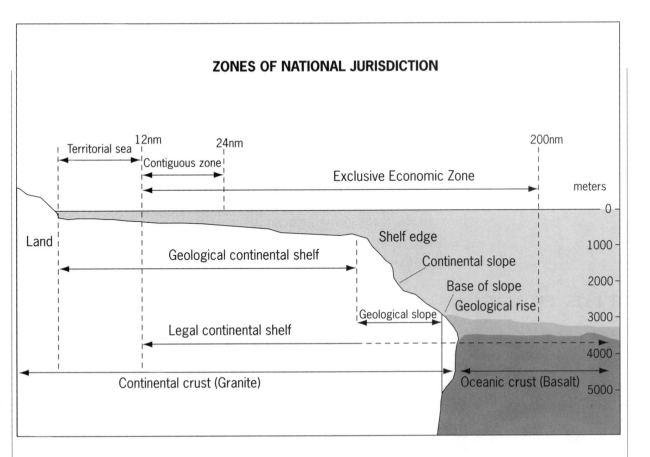

ZONES OF NATIONAL JURISDICTION

TRUMAN'S BOMBSHELL

As was so often the case during most of the twentieth century, superpower competition and unilateral action did much to destroy a generally beneficial, emerging accommodation among coastal countries. The Soviets must take part of the blame: in 1927 they expanded their territorial sea claim to 12 nautical miles and urged other countries to do the same.

But it was the United States that set off the time bomb. On September 28, 1945, President Harry Truman issued a proclamation that claimed for the United States all the resources that might lie upon and beneath the continental shelf—the floor of the shallow ocean immediately adjacent to our coasts, to a depth of 100 fathoms (about 660 feet).

What led the President to make such a sweeping claim? The

answer is resources—new resources known or believed to lie on or beneath the ocean floor. It was becoming clear to scientists that the submerged margins of the continents are made of the same rocks as the landmasses on which we live, and thus would contain the same raw materials, including oil. New technologies were bringing these resources within reach. Imagine the potential effect of the three-mile territorial sea, to which the United States still subscribed: just beyond the three-mile limit, some foreign country could legally set up oil extraction platforms. The Truman Proclamation forestalled such a situation.

THE RUSH IS ON

Almost immediately after the Truman Proclamation was issued, countries responded by matching or exceeding its terms. Mexico and Argentina claimed not only the continental shelf and its resources, but also the waters above. Chile and Peru, countries that have no extensive continental shelves, but are major fishing States, argued that they were unfairly disadvantaged and extended their territorial sea to 200 nautical miles from their coasts! Other countries, some egged on by the Soviet Union, announced various claims to adjacent waters and submerged lands. Some international agreement was obviously needed, and the United Nations sprang into action.

Thus was born the acronym UNCLOS—United Nations Conferences on the Law of the Sea. The first of these was convened in 1958, another in 1960, and a third in 1973; the third UNCLOS lasted through 11 sessions until 1982.

Out of these conferences came a host of international conventions relating to the world's seas and oceans, among which are the following:

- States may claim a maximum of 12 nautical miles of territorial sea.
- States may claim a 200 nautical mile Exclusive Economic Zone (EEZ), measured from their coastlines or baselines, where the coastal states have "sovereign rights for the purpose of exploring,

conserving, and managing the natural resources."

- States have rights over resources on and beneath the continental shelf, defined as "the seabed and subsoil of the submarine areas that extend beyond its territorial sea throughout the natural prolongation of its land territory to the outer edge of the continental margin, or to a distance of 200 nautical miles."

UNCLOS III gathered steam during President Jimmy Carter's Administration, and the U.S. delegation to the conference played a considerable role in writing the proposed treaties. But after Ronald Reagan took over the White House, he instructed the U.S. representatives not to sign the document—and only three other states joined in that refusal. Why would the United States not ratify a treaty it shared in developing and 154 other countries signed?

COMMON HERITAGE OF MANKIND

The major American objection to the treaty (which consists of 320 articles and 9 annexes on issues ranging from navigation and fishing rights to scientific research and environmental protection) centers on an underlying premise. The premise is that what remains of the high seas and the ocean floor beneath them are part of a "common heritage of mankind" that must be shared under the terms of "a just and equitable economic order." Section 11 of the treaty would create a global organization to permit developing nations to participate in and reap the benefits of the exploitation of whatever lies in and under the high seas, including minerals such as nickel, manganese, cobalt, and copper, all of which have been found in recoverable amounts on the ocean floor.

Developing countries applauded this section of the treaty, and were supported by the (then) Soviet Union. But the United States position after 1980 was that the treaty imposed unacceptable limitations on American efforts and capacities to mine strategic minerals wherever they were to be found. Further, the United States argued, the technology to do the mining would come exclusively from the

CHOKE POINTS

Before and during Desert Storm, ships under allied flags sailed into the Persian Gulf from the Gulf of Oman through one of the world's many *choke points*: the Strait of Hormuz. Between 65 and 90 miles (104 and 145 kilometers) wide, the Strait of Hormuz separates Iran and Oman. Small islands and shallows confine traversing ships to a narrow channel. They must slow down and twist and turn along a predictable course. In short, they are vulnerable.

The Strait of Hormuz is one of as many as a hundred such choke points: other famous ones are Gibraltar, the Bosporus, the Strait of Malacca, and the Strait of Magellan. In all these, territorial seas meet or overlap, and navigational channels often compel ships to cross into and out of the waters of the adjacent states.

Customary international law guarantees freedom of transit for the ships of all countries through these narrow channels (or canals: some are artificial). But they're not called choke points for nothing. Before the Camp David accord, Egypt would not let Israeli ships through the Suez Canal. Islands in the Strait of Magellan have been claimed by both Chile and Argentina. Pirates stalk vessels traversing the Strait of Malacca.

The world's choke points are always potential hot spots, places where congestion, delays, accidents, and aggressiveness can lead to wider conflict. The international free passage guaranteed by UNCLOS may occasionally be violated, but overall this is one of the United Nation's most notable achievements.

developed countries; the beneficiaries were exclusively the developing countries. And last, the global "collective" that would control this common heritage of mankind smacked of a Soviet-style operation at a time when the world was moving toward freer markets.

But the U.S. opposition had a price. While Section 11 was not easy to swallow in Washington, other provisions of the treaty were highly favorable to the United States, including the right to hold military maneuvers on the remaining high seas and the right of free passage through numerous strategic straits and other sea routes. And since countries are not strictly bound by the treaty now, some have begun to put restrictions on passage through their territorial waters, on fishing, and on other maritime activities. To forestall any more of this, the treaty must become part of international law. In 1993, President Bill Clinton's Administration began an effort to change the seabed mining provision before ratification, to prevent the treaty from passing into law with the common heritage of mankind section intact.

EXCLUSIVE ECONOMIC ZONE

With the adoption of a 200-mile Exclusive Economic Zone (EEZ), the world map changed overnight. Even the 12-mile territorial sea was a relatively narrow zone of offshore water; 200 miles, however, made quite a difference. Within the EEZs of the world lie virtually all the offshore oil and gas reserves, most of the recoverable metals, and the great majority of other maritime resources presently within reach of technology. And, as the map shows, the majority of the countries benefiting from large EEZs are already well-off, including the United States, Japan, Canada, Australia, and New Zealand.

The maritime map of the world now shows a large number of circular EEZs around islands, notably in the Pacific Ocean, greatly diminishing the remaining high seas. Consider this: a small island, just one square mile in area, has an EEZ extending 200 nautical miles (230 statute miles) in all directions. The country possessing such an island can exploit the resources in that vast EEZ or sell or lease such resources. The 200-mile EEZ suddenly made islands more valuable than ever. That raises the question:

WHAT IS AN ISLAND?

That should be a geographic trivia question with an easy answer: an island is a piece of land completely surrounded by water. But in a world of territorial seas and Exclusive Economic Zones, this is not good enough. Based on this answer, there are more than a half million islands on the planet. Even a small island yields a 166,000 square mile EEZ. If all of them were so endowed, there would be little left of the high seas.

The UNCLOS conferees, therefore, came up with a more specific definition of an island: "a naturally formed area of land, surrounded by water, which is above water at high tide." That took care of rocks and sandbanks that stick up above the waves only at low tide, but it still left far too many specks of rock and sand for which owner-governments might demand territorial seas and EEZs. So the definition got an addendum. "Rocks which cannot sustain human habitation or economic life of their own shall have no exclusive economic zone or continental shelf." Read that statement carefully, and you will note that they apparently *are* eligible for territorial seas. This has encouraged island-owners to establish these, against the obvious intent of the treaty. There's more to a small island than meets the eye.

ROCKALL

Have a look at a map of the North Atlantic Ocean. Well over 200 miles (320 kilometers) west of Scotland's westernmost islands (the Hebrides) lies an aptly named island: Rockall (it's all rock). Rockall stands above the water at high tide: if it gets its own EEZ, its zone will meet that drawn off the Hebrides, pushing Britain's maritime border hundreds of miles westward into the Atlantic. The British saw the opportunity coming: no one had claimed Rockall until they did, as recently as 1955. But can Rockall sustain human life? Well, put a lighthouse and a lighthousekeeper on it....

THE SPRATLY ISLANDS

Here's a more serious problem: the 200 or so Spratly Islands in the middle of the South China Sea. At last count, all or some of the Spratlys were claimed by China, Vietnam, the Philippines, Malaysia, Taiwan, and even Brunei.

For centuries no one had claimed the Spratlys, but the actual and potential value of islands has grown exponentially of late. In the case of the Spratlys, which lie outside the EEZs of all the claimants (let alone in territorial seas), that value involves oil. The shallow zones of the South China Sea may contain huge oil reserves, and the way to get hold of them is to claim islands—and then to draw 200-nautical-mile EEZs around them.

The Spratlys are not the only targets of maritime competition in the South China Sea. To the northwest lie the Paracel Islands, long held by Vietnam—until 1974, when China ousted the Vietnamese by force.

When I visited Vietnam in 1993, a geography teacher at a Saigon high school (I know it's Ho Chi Minh City now, but people there still call it Saigon) talked to me about Vietnam's relationships with other countries. "Yes, we have unfinished problems with the U.S.A.," he said. "You pursue the missing-in-action issue and we don't like the embargo you put on us. But in the end, we didn't lose any territory to you. It was the Chinese who fought us afterward and took our land. The Paracels are ours, and we want them back, and that goes for the fishing grounds we've always used."

Small islands can generate huge conflicts, as the South China Sea will remind us.

THE KURILES

Japan and the former Soviet Union never signed a peace treaty to end their Second World War conflict. The reason? There are actually four, all of them on the map just to the northeast of Japan's northernmost large island, Hokkaido. Their names are Habomai, Shikotan, Kunashiri, and Etorofu. The Japanese call these rocky island groups of the Kurile Island chain their Northern Territories. The Soviets occupied them late in the war, but they never gave them back to Japan. Now they are part of Russia, and the Russians have not turned them over, either.

The islands themselves are no great prize. During the Second World War, the Japanese brought 40,000 forced laborers, most of them Koreans, to the islands to mine the minerals found there. When the Red Army overran them in 1945, the Japanese were ordered out, and most of the Koreans fled. Today, the population of about 50,000 is mostly Russian, many of them members of the military based on the islands and their families. At their closest point, the islands are only 3 miles (5 kilometers) from Japanese soil, a constant and visible reminder of Japan's defeat and the loss of

Facing Page: Since the discovery of oil and natural gas in the 1950s, European countries facing the North Sea have allocated sectors of the seafloor to themselves.

Japanese land. Moreover, their territorial waters bring Russia even closer, so the islands' geostrategic importance far exceeds their economic potential.

Attempts to settle the issue have failed. In 1956, Moscow offered to return the tiniest two, Shikotan and Habomai, but the Japanese refused, demanding all four islands back. In 1989, Soviet President Mikhail Gorbachev visited Tokyo in the hope of securing an agreement. The Japanese, it was widely reported, offered an aid and development package worth $26 billion U.S. to develop Russia's eastern zone—its Pacific Rim and the vast resources of the eastern Siberian interior. This would have begun the transformation of Russia's Far East, stimulated the ports of Nakhodka and Vladivostok, and made Russia a participant in the spectacular growth of the Austrasian Pacific Rim.

But it was not to be. Subsequently, Russian President Boris Yeltsin was also unable to come to terms with Japan on this issue, facing opposition from the islands' inhabitants and from his own government in Moscow. And so the Second World War, a half century after its conclusion, continues to cast a shadow over this northernmost segment of the western Pacific Rim.

DIVIDED SEAS

When the UNCLOS conferees agreed on the notion of a wide EEZ, they knew full well that they were, in effect, parceling out many of the world's smaller seas among coastal countries. Many seas are much less than 400 nautical miles wide, including the North Sea, the Mediterranean, and the Caribbean.

Across these seas, maritime boundaries were drawn that made neighbors out of formerly remote countries: Norway and the United Kingdom, Italy and Libya, the United States and Venezuela.

These maritime boundaries divide not only the sea, but also what lies beneath it, and as fate would have it, many of the

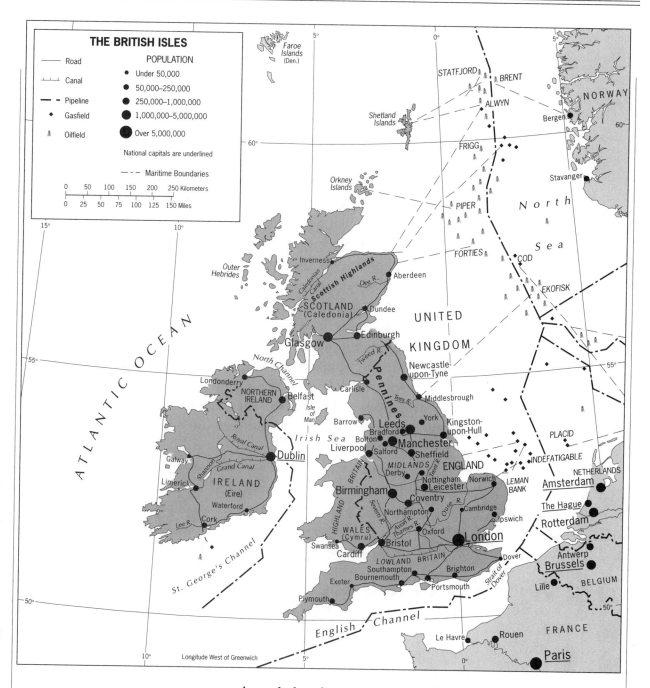

THE BRITISH ISLES

Road
Canal
Pipeline
Gasfield
Oilfield

POPULATION
- Under 50,000
- 50,000–250,000
- 250,000–1,000,000
- 1,000,000–5,000,000
- Over 5,000,000

National capitals are underlined

Maritime Boundaries

0   50   100 150 200 250 Kilometers
0   25   50   75 100 125 150 Miles

resources down below lie right on these borders. There's probably
no better example in the world than the North Sea boundaries
between Britain, the Netherlands, Germany, Denmark, and Norway.

As the map on page 183 shows, the belt of oil and gas reserves found under the North Sea follows almost exactly the maritime political boundary. When you sail across this part of the North Sea, the border zone is a marine beehive of activity: oil platforms, ships, helicopters, and other equipment and activity all mark the spot. There's no land in sight, but the area is as busy as any place shoreside. And here's the old boundary problem again: some oil and gas reserves actually straddle the border. How much did nature put on one side of it, and how much on the other? That's a critical question, especially for Norway, whose economy has come to depend heavily on North Sea oil from beneath its EEZ.

With all its tiny islands and numerous nations, the Caribbean's maritime boundaries are more intricate even than the North Sea's. But no major oil reserves have yet been found astride any Caribbean boundaries, so that sea has been allocated with little dispute. The Caribbean hosts the world's largest fleet of cruise ships, and the territorial seas and EEZs of coastal and island Caribbean countries are used to extract fees from cruise line companies. Otherwise, the Caribbean's water boundaries are as yet without economic significance. The Gulf of Mexico, on the other hand, is divided between the United States and Mexico—and this marine region *has* yielded important energy resources. There, the future may be quite different.

THE WORLD LAKE CONCEPT

The successive UNCLOS conferences constituted a major international achievement: they controlled the scramble for the oceans and contained the expansion of maritime claims, albeit at the cost of awarding a wide EEZ to all coastal states. Here was an early sign of the New World Order we are supposed to anticipate, and it seemed to work with the exception only of the "common heritage of mankind" clause.

But international agreements are made to be broken, and if the New World Order fails, then the next step in maritime aggrandizement may not be far behind.

What would happen if all the waters of the world were divided as the North, Mediterranean, and Caribbean Seas (among others) are today? Maps showing where the world's oceanic boundaries would lie began to make their appearance during the 1970s, and it is interesting to see what may lie in store.

The map shows how much of the world's oceans has already been allocated under the EEZ clause: huge circular zones, over 460 statute miles (740 kilometers) in diameter, surround islands throughout the Atlantic, Pacific, and Indian Oceans.

But look at the maritime domains if the oceans were divided in accordance with UNCLOS median-line rules. In this era of decolonization, many states still own distant islands; as a result, much of the South Atlantic would belong to the United Kingdom. Portugal, by virtue of its control over the Azores, becomes a neighbor of Canada in the North Atlantic!

In the Pacific, the United States is the big potential winner and Canada the loser, squeezed as it is between Alaska and Washington State. In the South Pacific, Chile and New Zealand gain superterritories. And in the Indian Ocean, Australia reigns supreme, although France (also strongly represented in the Pacific) has large potential claims.

What the map represents may not come to pass, but in the smaller seas of the world, this has already happened. Today, the world's attention is elsewhere, technologies of mineral recovery are still improving, and mineral availability makes maritime recovery a generally uneconomic proposition (except for oil and gas). But times change, resources do become scarce, prices can and do rise. Then the pressures on the oceans will increase, and self-interested countries may turn again to the world lake concept to justify expanded claims to our last frontier.

ANTARCTICA

Even as EEZs reduce the high seas, another geographic realm faces the threat of rising international competition: Antarctica and the Southern Ocean.

Antarctica is the world's most remote continent, and its environments are the harshest on Earth. The region has been described as the "white desert" and as the "home of the blizzard." The Southern Ocean's waters are frigid, filled with icebergs, and whipped by raging storms.

Yet the region has attracted international attention, first by whalers and sealers in the Southern Ocean, then by explorers planting national flags on the icy mainland. Reaching the South Pole became an obsession that motivated scientific societies and individual adventurers to make overland journeys into the interior. A Norwegian, Roald Amundsen, reached the pole first, in 1911; the British explorer Robert Falcon Scott found the Norwegian flag when he reached the same point just a month later.

Such explorations and heroics led to national claims. Coastal claims on the Antarctic perimeter were extended to the pole as pie-shaped sectors of Antarctica. Australia's was the largest, followed by the claims of Norway and New Zealand. Inevitably, some claims overlapped. Where Antarctica extends (in the Palmer Peninsula) toward South America, the claims of three states (Argentina, Chile, and the United Kingdom) clash. France has a narrow sector, and a large area named Marie Byrd Land remains unclaimed.

After the Second World War, scientific research on the southern region expanded. States recognized the need for international cooperation, and in 1959, a group of 13 signed and ratified the Antarctic Treaty, including all seven claimant states. The treaty took effect in 1961 and was to remain in effect for 30 years. Under its chief provisions, all states agreed to reserve Antarctica for peaceful, scientific purposes; territorial claims were held in abeyance, and no new claims were recognized.

Significantly, the Antarctic Treaty did not address resource exploitation. Fishing for krill and other marine life in the Southern Ocean intensified. During the global energy crisis of the 1970s, interest in the mineral and fuel potential of the frozen continent and its deeply submerged continental shelves also grew. As technology brings more such raw materials within reach, the southern realm will attract more attention.

The Antarctic Treaty was reconstrued as the Wellington Agreement in 1991, and there has been growing concern that it might not be strong enough. In particular, it does not settle the question of resource exploitation.

In an age of growing national self-interest and increasing resource needs, the possibility exists that international rivalry in the Southern realm will intensify and produce a confrontation that the treaty cannot prevent. The Southern realm is truly the globe's last frontier, and the partition of its lands and waters is a process fraught with dangers.

C H A P T E R   1 0

STATE OF IRRELEVANCE?

Every year, the Bureau of Intelligence and Research, an office in the U.S. Department of State, publishes a list of what it calls the Nations of the World. The list has grown longer, especially over the past half decade. Some unfamiliar "nations" have joined it: Moldova, Eritrea, Belarus. But you will look in vain for some other, very old and very large nations. The Kurds, 20 million strong, are not there. Nor are the 8 million Zulus or the 6 million Palestinians.

So what's wrong with this list? It's a matter of geography: the misuse of a term. The list is not a list of the nations of the world, but of the states. There is a crucial difference. Some states incorporate more than one nation. Other nations spill over into neighboring states: there are millions of Hungarians outside Hungary. And some nations exist without statehood, as do the Kurds and—perhaps less permanently—the Palestinians.

All this would matter less if the concept of nation were merely an academic abstraction. But in fact, the idea of nation can arouse passions as strong as those provoked by the idea of race. I have heard the words "we are a nation" spoken fervently by Scots, Catalunians, native Hawai'ians, Croatians, Ibos, Flemish, and many others. For many of us, nationhood is a routine fact of life. For many others, it is a constant, elusive goal.

The neutral word, of course, is "country," a word of Latin root so old that its original meaning—the land opposite—has fortunately faded. So there is something contrary about "country" too, but nothing like "nation."

NATION-STATE

How did this conundrum arise? The answer lies at the source of the modern state, in Europe. Between 1650 and 1850, Europe was transformed by three massive revolutions: in agriculture, industry, and government. The agrarian revolution enhanced farm productivity and created unprecedented surpluses of food. The Industrial Revolution mobilized manufacturing and stimulated urbanization. And the political revolution ended the era of omnipotent royalty and ushered in the age of representative government.

The political revolution created the first modern states and consolidated the map of Europe. The process is not over, as the crisis in Yugoslavia underscores, but out of the revolutionary period emerged something the Europeans called the "national state," or nation-state. In the national state, nation and territory were coterminous, or very nearly so. Following its murderous revolution, France became a prototype of this nation-state. The British on their island, and the Germans between the Rhine and the shores of the Baltic, also consolidated into nation-states.

In this European model, states were the domains of more or less homogeneous nations, and citizens of such states acquired legal nationality. Spain, Portugal, Italy, the Netherlands, Sweden, Norway, and Denmark were among early territorial states in which the overwhelming majority of the inhabitants were, and regarded themselves as being, members of nations as well as citizens of states. For these people, nationality meant the same as "subject of the state." And so the notion of the nation-state became the prevailing European model.

But not everywhere. Belgium, as noted earlier, emerged as a state of two nations, not one. Venerable Switzerland, whose origins predate Europe's revolutions by many centuries, is a state of three nations and several smaller minorities. And when, after the First World War, Europeans redrew the map of Eastern Europe, they forgot the European model and combined several nations into one

THE KURDS: A NATION WITHOUT A STATE

Political maps of the region do not show it, but where Turkey, Iraq, and Iran meet, the cultural landscape is not Turkish, Iraqi, or Iranian. Here live the Kurds, a factious and fragmented nation of about 24 million (no certainty exists about their exact numbers). More Kurds live in Turkey than in any other country—perhaps as many as 10 million; possibly as many as 7 million live in Iran, 3 million in Iraq, and smaller numbers in Syria, Armenia, and even Azerbaijan.

The Kurds have lived in this isolated, mountainous frontier zone for more than 2,000 years. They may be a nation, but they have no state; nor do they have the international exposure that people of other stateless nations, such as the Palestinians, receive. Baghdad's brutal repression of the Iraqi Kurds created more than 2 million refugees in 1991 and made world headlines—but soon the world forgot again.

The Kurds have been pawns in an endless series of geopolitical struggles. The Shah of Iran encouraged their rebellion against Iraq during a quarrel in the 1970s, then abandoned them to retribution when the conflict ended. The Turks as recently as the 1980s forbade the public use of Kurdish speech in the Kurdish area of southeastern Turkey. And the Kurds have often engaged in warfare among themselves.

Many Kurds dream of a day when their fractured homeland will become a nation-state. Most would agree that the city of Diyarbakir, now in Turkey, would become the capital; it is the closest any Kurdish town comes to a dominant city. After the 1991 Gulf War, the United Nations established a security zone for Kurds in Iraq, extending from the 36th parallel northward to the borders with Turkey and Iran. This was done to encourage the refugees to return and to protect them against further Iraqi mistreatment. But this also may be the closest the Kurds will come, in the foreseeable future, to their dream of a free Kurdistan. The stakes—territory, oil, farmland, river sources—are too great for the states in this immediate area to agree to move their boundaries.

uneasy union that was destined to break apart.

The synonymous use of state and nation, therefore, has European roots. That's why we still refer to our nationality when we identify the country we live in, even when we're Walloons living in Belgium, Basques living in Spain, Quebecois living in Canada, or Ibos living in Nigeria. And we continue to refer erroneously to multinational countries as nations, when the simple geographic term states would put it right.

EXPORTING THE MODEL

Anthropologists report that processes of state formation occurred independently in widely separated parts of the world, beginning soon after organized agriculture made possible the storing, control, and distribution of surpluses—thus creating social layers and classes. The European nation-state model was a modern version of this experimentation, which has been going on for thousands of years. But it was the European model, not its many predecessors, that was exported to all parts of the world. As it happened, the maturing of the European nation-state coincided with the expansion of European power around the globe. On the wings of colonialism, European ideas about statehood spread far and wide.

By the time that happened, during the eighteenth and nineteenth centuries, European philosophies about the state had already begun to diverge. Most of Europe's nation-states were highly centralized, with a powerful government; these *unitary* states, some still ruled by royalty, included the United Kingdom, France, Spain, Germany, the Netherlands, and the Scandinavian countries. But from Colonial European communities came notions of regional-cooperative government, in which some of the powers of the center were assigned to the provinces. Switzerland already possessed this kind of federal government, and, so the Europeans thought, this might well fit the cultural diversity of colonies.

Whether unitary or federal, the European models of the state had this in common: they possessed a defined portion of the Earth's surface and were therefore separated by boundaries (not frontiers) from their neighbors; they contained one or more of Europe's nations; they had a governmental system centered in a capital city; and they were served by infrastructures laid out to promote economic development and national power. In an age of colonial competition abroad and intermittent war at home, power was a key ingredient of the successful state.

A German political geographer, Friedrich Ratzel (1844–1904), likened the evolving European nation-state to a biological organism. He argued that the nation, being an aggregate of living beings, would mirror the life cycle of an individual: it would have to be nourished by absorbing other cultures and by expanding into other lands. Colonial acquisition, he suggested, was good for states; defined and delimited boundaries were bad for it. States should be able to vie for territory; when that option was closed, the state, like an old person, would become senile, and the nation would wither. He based this organic theory of the state on historical analysis, and he published his opinions in scholarly journals. But it was not long before these ideas found their way, via his students, into German politics. Later, they became part of Nazi ideology, promoted especially by a strategist named Karl Haushofer. His brand of *geopolitik* gave that term such a bad name that it went into disuse by geographers for several decades.

In the meantime, the European powers were playing power politics on the world stage, laying out, in large measure, the boundary framework with which the world has been saddled ever since. The colonists drew borders where there had been none, built capitals where none had existed, and constructed facilities to enable them to exploit and administer their overseas domains. Little did they imagine, in the nineteenth century, that the lands they were organizing would someday be on the list of nations of the world.

Nor were colonial domains laid out with the interests of indigenous peoples in mind. Take the case of Africa. As the colonial map of Africa got more crowded, the colonial powers decided to meet and divide the continent rather than fight over it. In 1884, they got together in Berlin to negotiate, and out of that infamous Berlin Conference came, essentially, the map of Africa as we know it today. Only Ethiopia (then known as Abyssinia) and Liberia survived the power grab—Ethiopia because of its natural mountain-fortress protection and Liberia because of its American connections. The rest of Africa was assigned to the British, French, Portuguese, Belgians, and Germans (Spain got a tiny corner, too). The colonial boundaries were drawn, in part, to place African peoples under a single colonial master. But often, the new boundaries divided African nations, split-

THE BERLIN CONFERENCE

In November 1884, the imperial chancellor and architect of the German Empire, Otto von Bismarck, convened a conference of 14 powerful states, including the United States, to settle the political partitioning of Africa. Bismarck not only wanted to expand German spheres of influence in Africa but also sought to play Germany's colonial rivals against one another to the Germans' advantage. The major colonial contestants in Africa were the British, who held beachheads along the West, South, and East African coasts; the French, whose main sphere of activity was in the area of the Senegal River and north of the Zaire (Congo) Basin; the Portuguese, who now desired to extend their coastal stations in Angola and Moçambique deep into the interior; King Leopold II of Belgium, who was amassing a personal domain in the Congo; and Germany itself, active in areas where the designs of other colonial powers might be obstructed, as in Togo (between British holdings), Cameroon (a wedge into French spheres), South West Africa (taken from under British noses in a swift strategic move), and East Africa (where the effect was to break the British design for a Cape-to-Cairo axis).

When the conference convened in Berlin, more than 80 percent of Africa was still under traditional African rule.

Nonetheless, the colonial powers' representatives drew their boundary lines across the entire map. These lines were drawn through known as well as unknown regions, pieces of territory were haggled over, boundaries were erased and redrawn, and sections of African real estate were exchanged in response to urgings from European governments. In the process, African peoples were divided, unified regions were ripped apart, hostile societies were thrown together, hinterlands were disrupted, and migration routes were closed off. All of this was not felt immediately, of course, but these were some of the effects when the colonial powers began to consolidate their holdings, and the boundaries on paper became barriers on the African landscape.

The Berlin Conference was Africa's undoing in more ways than one. The colonial powers superimposed their domains on the African continent; when independence returned to Africa after 1950, the realm had by then acquired a legacy of political fragmentation that could neither be eliminated nor made to operate satisfactorily. The African politico-geographic map is thus a permanent liability that resulted from three months of ignorant, greedy acquisitiveness during the period of Europe's insatiable search for minerals and markets.

ting them between French and British domains, or German and British, or Belgian and Portuguese.

Next, the colonial powers established internal divisions within their vast empires, creating units of administrative convenience and splitting African peoples still further, or combining peoples with adversarial histories. Just a glance at the African map today reveals the capriciousness involved: look at the Belgian Congo (now Zaire), with its huge interior bulk and tiny corridor to the sea, or German South West Africa (now

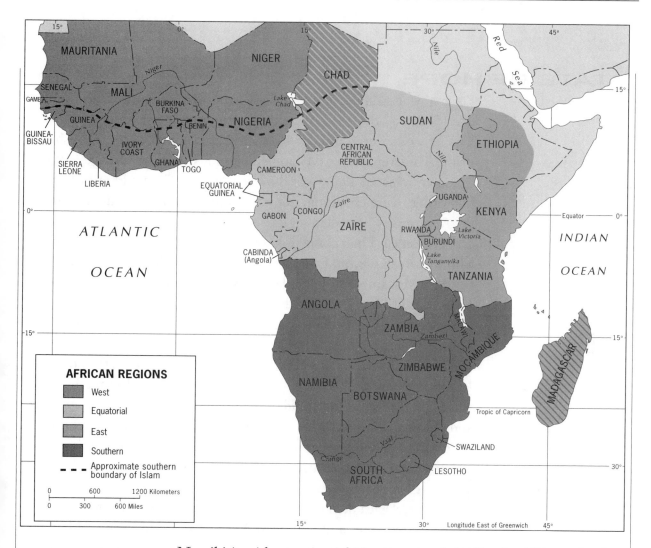

Namibia), with an arm sticking out toward the Zambezi River. Or the strips of land named Togo and Benin. Or the landlocked isolation of Burkina Faso, Mali, Niger, and Chad—all parts of the former French West African empire.

Africa was not alone in this. The British tried to rule their South Asian empire as a multicultural whole, while the French partitioned their Southeast Asian colony as they did in Africa. In short, the world's international boundary framework was anything but ready, or suitable, when the wave of decolonization arose to sweep the colonies toward independence and statehood.

STATE OF HOPE

For many academics, and especially geographers, the events of the late 1950s and the 1960s were exciting, sometimes breathtaking. The idea of sovereignty was sweeping the colonial world. British Prime Minister Harold Macmillan's famous "wind of change" speech, which anticipated a new world of free "nations," rang in our ears. Old orders were changing, not only in the colonial world but also in the Americas, where democracy seemed to be stirring. At Northwestern, I had come to know and admire Eduardo Mondlane, who we all knew would be president or prime minister of an independent Mozambique some day. As a member of an ongoing seminar chaired by the famed anthropologist Melville Herskovits, I had the opportunity to listen to the leaders of the new Africa: Sekou Toure of Guinea, Tom Mboya of Kenya, and others. With leaders like these, and others not yet free to lead their countries (Jomo Kenyatta, Kwame Nkrumah, Julius Nyerere, Patrice Lumumba), Africa's future as an independent realm appeared bright. Economists and political scientists who appeared before the Herskovits seminar proclaimed that a new era of African prosperity was around the corner, the remaining colonial yoke being the only obstacle. And that yoke was being lifted.

To me, Africa held center stage, but the wind of change also swept across South and Southeast Asia and the Caribbean. What amazed me was that the world could change so much, so fast, with relatively little conflict. In Africa, the Mau Mau rebellion in Kenya, a relatively minor but cruel skirmish, was the largest contest. India, Pakistan, and Sri Lanka had emerged following the creation of new boundaries and much human dislocation, but the British were out without a war. The Dutch struggle for continued control in Indonesia had been short-lived. The French were being ousted from Indochina in a classic late colonial war, and Algeria was in revolution. But negotiated withdrawal rather than all-out combat seemed to be the norm. I envisaged a new world order in which the former colonies, now sovereign democracies, would join the European states on a plane of political equals.

So did many of my colleagues. At Michigan State, I helped train the first contingent of Peace Corps volunteers and spent a summer learning Swahili. The University set up a collaborative venture with Nigeria, creating at Nsukka the University of Nigeria; I hurried over there to lecture at the first opportunity. Periodically, the *Huntley- Brinkley Report* would announce the independence of yet another African state, and I plotted ways to secure research funds to enable me to see it for myself. Once in Africa, I kept running into colleagues on the same mission, performing research and writing about the newly emerging African states. The political scientist David Apter wrote glowingly about the prospects for Ghana, endowed at independence with an electoral system based on the Westminster model—and with ample cash reserves to launch its new era. The historian James Hooker anticipated that the wind of change would be slowed down in Southern Africa but that Rhodesia would be independent by 1970 and *apartheid* abolished within 10 years after that. Most of all, we admired so many Africans who had come out of the era of colonialism with so little rancor. True, there were dreadful scenes in the Belgian Congo where white settlers in interior towns, isolated by the rapid course of events, were set upon and murdered by African mobs. But there was little sympathy for the settlers. The greater goal—a free, thriving, democratic, stable Africa—obscured all else, and to this, it seemed, the vast majority of Africans were committed.

STATE OF SHOCK

Ten years later, most of our hopes had been dashed, and much of Africa was neither free nor stable. Mondlane had been assassinated, by Portuguese agents it was rumored, in Tanzania. Lumumba had been killed during the power struggle for the Congo. Mboya had been shot down on a Nairobi street. Nkrumah survived several assassination attempts but was in exile, his country's coffers bare and its electoral system in a shambles. A war of secession brought the name Biafra to world headlines and cost a million Nigerians their lives. In the south, revolutionary wars with enormous costs in lives and infrastructures raged in Mozambique and Angola, and Rhodesia had declared itself independent—not the African majority there, but the white minority. The result was a prolonged and dreadful civil war, and when it was over, a further conflict began between the two most powerful African nations within the new Zimbabwe. As for South Africa, its rulers proceeded with a new grand design, creating a set of 10 quasi-independent African homelands under a program called Separate Development.

Now the news from Africa was carrying stories of excess and failure. Uganda descended into an abyss of destruction during the regime of Idi Amin. In the Central African Republic, Emperor Bokassa ruled by terror. Zaire was becoming a dictatorship. Military coup followed military coup in countries from Guinea to Madagascar. Scholars were discussing the "failure of African leadership."

It was a shocking reversal of fortunes, and there were many underlying causes. One of these was the failure of the European state system in African settings. Belatedly and hurriedly, the departing colonial powers sought to establish European state systems in their dependencies: the French on the unitary, all-roads-lead-to-Paris model; the British on more decentralized models when circumstances seemed to demand it (Nigeria became independent as a federation). But these systems did not have the time to mature, and, in any case, the challenges they confronted, created in part by the Europeans' own boundary mosaic, were well-nigh insurmountable. A few notable successes (Senegal, Ivory Coast, Kenya, Tanzania, postwar Zimbabwe) were soon outnumbered by failures in dozens of other countries, ranging from Guinea and Sierra Leone to Mozambique and Madagascar. Even after overcoming civil war and restructuring its federal system several times, Nigeria, too, succumbed to military dictatorship.

The tragedy of Rwanda and Burundi, two small but populous countries wedged between Zaire and Tanzania, landlocked, poor, and strife-ridden, underscores the weaknesses of the African state. Here the cattle-owning Tutsi (Watusi in earlier parlance) met the crop-farming Hutu (Bahutu) centuries ago and, though in the minority, dominated the farmers throughout their domain. The European intrusion served to consolidate the Tutsi advantage and bequeathed their territories with state boundaries, borders awarded to the Belgians for fighting the Germans in East Africa during the First World War. The Belgians exploited their holdings for the labor force they contained, and countless thousands of Hutu as well as Tutsi worked in the mines of the Congo.

Back home, the distinction between Tutsi and Hutu became less an ethnic and more an economic one. While the Tutsi continued to organize themselves in kingdoms, hoarding the wealth, in the form of cattle and land, and the Hutu tilled the soil, there was intermarriage and considerable transculturation. To be a Tutsi became more a matter of status, less a matter of ethnicity.

When Belgium's dependency was split into two sovereign states, in 1962, and Belgian authority ended, it was a matter of time before score-settling began in Rwanda and Burundi. Long before the horrifying events of 1993 and 1994, hundreds of thousands had died in a series of civil conflicts. In 1972, an estimated 100,000 Hutu were slaughtered in Burundi. In Rwanda, meanwhile, Hutu went on a rampage that sent more than 100,000 Tutsi into exile in Uganda and Zaire. This fateful event laid the foundation for the invasion of 1994, when a Tutsi force trained in Uganda forced its way to Kigali and took control of the country.

The collapse of Rwanda's fragile order in 1994 followed months of intermittent fighting, and led to one of the worst human tragedies of the twentieth century. Probably as many as 1 million of Rwanda's 8 million people were killed, many massacred in systematic fashion by Hutu militias obviously long-prepared for the purpose. Massive, desperate emigration streams carried more than 2 million Rwandans across the borders into Tanzania and Zaire. In the refugee camps, hunger and disease took their toll. In the meantime, the Tutsi-led Rwanda Patriotic Front forces, from their Ugandan exile, capitalized on the chaos to seize control of the country, setting the state for the next round of score-settling.

In Liberia and Somalia (and a growing number of other post-colonial African countries), the state is simply irrelevant in the context of such intense conflict; the real authority lies with local rulers and people who call themselves generals but who are elsewhere dubbed warlords. There is a Rwanda on the map, and it has a capital, secondary towns, roads, and boundaries. But the real Rwanda is not the one laid out by the Europeans to function as a national state. It never had what it would have needed to survive.

Of course Africa was not alone in this. Federal Pakistan, that awkward linkage between what is now Pakistan and Bangladesh, was another ill-fated British experiment. European-style democracy has not exactly been the hallmark of the former Dutch colony, Indonesia. And military dictatorships and other forms of autocratic rule have not been strangers to Middle or South America.

In short, the European state model was exported to the rest of the world, but in many cases it failed: the very boundary frameworks left behind by the colonial intervention virtually assured this failure. Every time I hear an American Secretary of State proclaim that regional problems must be solved "within existing boundaries," I wince; we should get used to the notion of changing boundaries rather than fighting over them. The future of world order and stability may depend upon a mechanism for such change.

The shocking descent of African and other postcolonial countries into disorder and, in some cases, chaos, should be seen against the geographic background. Many of these countries incorporated two, three, or more nations and faced problems far more severe than those confronted by the European model. The colonial countries had been given boundaries and capitals, but infrastructures were weak or nonexistent (Somalia is a prominent example of the latter) and national integration had barely begun. Further, they were enmeshed in an international economic system in which they had little or no say, and poverty overtook them. Political progress tends to occur when economic conditions are favorable, when economies decline, and the political fabric weakens. Europe today is a good example of this: heady advances toward a European Union were made while Europe's economies thrived. Now that Europe is having its own economic malaise, the Union's future does not look quite so bright.

Seen in this context, the very survival and continuity of many of the postcolonial states constitute something of a miracle. The endurance of India is such a marvel: a multicultural state of nearly 1 billion people, and a democracy at that, it has survived against all odds. Despite its civil war, Nigeria's continuity is remarkable: Nigeria incorporates three major nations and many smaller, yet culturally distinct, peoples. In the mid-1990s, Nigeria faced a growing religious challenge: the numerically dominant north is Islamic, while the south is mainly Christian and Animist. In this era of religious fundamentalism and cultural challenge, Nigeria's future is clouded, and again economics plays its role. Nigeria's oil wealth brought temporary prosperity and much borrowing; now the price of oil is down and the means to grease the political machine have receded.

Still, there are other successes, ones that don't make the nightly news. Somehow, despite a misguided foray into an enforced villagization program, Tanzania held together. Notwithstanding declining prices for its exports and of a withering, 10-year drought, Zambia managed a democratic political transition. Sparsely populated Botswana remains a democracy; Namibia escaped from South Africa's clutches in one piece.

And the great success story—although the story is not finished— is the relatively peaceful transition of power now in progress in South

Africa, the realm's richest country, the land where apartheid ruled, and where the Africa of the past and the Africa of the future came face to face in one last confrontation. That this confrontation remained essentially confined to politics, and did not precipitate the civil war so many observers expected, was a marvel and a triumph. When the great majority of Afrikaners, whose people and party were the architects of apartheid, followed their President de Klerk in dismantling not only that system but also their republic, a miracle unfolded. And when Nelson Mandela managed to channel the frustrations, resentments, suspicions, and divisions of his people toward democratic political goals, he personified a triumph of the human spirit. Some say that the April 1994 election was manipulated; that it was rigged to give Buthelezi's Inkatha movement just enough power to prevail in Natal, and the Afrikaner-Coloured alliance just enough support to stay in control at the Cape; and that the failure of the African National Congress to gain a two-thirds majority was designed to allay the fears of minorities, principally the whites. No matter: South Africa crossed the Rubicon, an achievement unmatched in the turbulent world of political transition.

South Africa still faces daunting problems. Its economy lies in a shambles made worse by the sanctions imposed against the apartheid regime. Hundreds of thousands of African youths, who were encouraged by the slogan "liberation before education" to leave their schools, now are unemployable and pose a social problem. The housing shortage is desperate. Crime is rampant and security is a nationwide concern. And there looms the question: How will South Africa manage after Mandela? In so many other African countries, the courage and charisma of those who led their states to independence proved to be all that welded their incipient nations together; once removed from the scene, that bond collapsed in a hurry. It is often said that in an emerging democracy it is the second election, not the first, that reveals the future. South Africa may prove the point.

Elsewhere in Africa, sadly, the picture is bleak. In poverty-stricken countries, the sole stake is power, and in the bounded state system now in worldwide use, power all too often resides in military cliques, elites, or dominant national groups. The creation of capital cities helped concentrate power geographically, to the advantage of those within and nearby and handicapping those in the country's periphery. When I asked a Somali colleague the leading cause of conflict in her country, she answered with one word: Mogadishu. The geographer David Mathew in his 1947 book on Ethiopia, describes Addis Ababa as "a mask, behind which the rest of Ethiopia hides." Fifty years later, his observation still rings true.

STATE OF TRANSITION

A few years ago I received, from a former graduate student now working in East Africa, a copy of the front page of the daily newspaper there. "Tribal War Breaks Out Again in Europe," said the headline. The article was about Yugoslavia, and in a related editorial, the paper wondered why such conflict is always referred to as tribal war in Africa, but "ethnic conflict" or some such in Europe.

The fact is that the European model of the state is failing, even in Europe itself. Yugoslavia is an extreme case (so far), but European countries are experiencing forces similar to those threatening to break up "nations" from Sri Lanka to Canada and from Cyprus to Angola. Geographers long ago termed this process *devolution*, the reverse of evolution, and it afflicts even the oldest and stablest of nation-states.

We can see the evidence on the map. Separatist movements are active in Spain (the Basques, the Catalunians), in France (Corsica), in Britain (Scotland, not to mention Northern Ireland), Italy (the Ancona Line looks more and more like a boundary between a future north and south), and Belgium (where Flanders wants out). Czechoslovakia has already fractured, and Yugoslavia is no more.

All this is happening at the very time Western Europe is trying to unify! That would seem to be a contradictory situation, but it isn't. Both the collaborative, so-called supranational ventures of the Europeans, mobilized during good economic times, and the secessionism exhibited by the Catalunians and others evince a fundamental dissatisfaction with the bounded-space, capital-dominated European nation-state model. It just so happens that the national governments seeking greater integration have greater resources at their disposal than the smaller groups seeking more autonomy from those very governments. In short: from Sweden to South Africa and from Cambodia to Canada, the nation-state system is coming up wanting.

What lies ahead? It's clearly impossible to envisage a world in which every identifiable cultural group will have its own state; in Africa alone, that would create more than 1,000 such entities. On the other hand, supranational endeavors run into endless difficulties, even when the richer countries of the world try to unify. The North American Free Trade Agreement (NAFTA) was far from an international merger, and look what it took to get it approved.

How far should national self-determination within existing states be allowed to go? Take the example of the embattled former Soviet Republic of Georgia. While the Georgians were trying to free themselves of Soviet domination, they spoke and wrote glowingly about the rights of nations to determine their own futures. But once the Georgian state was free of Moscow, those same Georgians suppressed the aspirations of their minority nations, the Ossetians and the Abkhazians, to form their own free states. I carefully monitored the international press during these events and concluded that while there was widespread support for the Georgian nationalists, the aspirations of the Ossetians and the Abkhazians were viewed much less favorably. (This may well be the result of lower geographic literacy when it comes to smaller nations. Ossetians and Abkhazians were even less familiar to the general public than Georgians were).

The question, therefore, is not easily answered. A poll taken in Europe indicated that there was much stronger support for

Catalunian autonomy than for Basque independence; the reason, in the words of one respondent, was that "Catalunia has a fairly large territory and a capital (Barcelona) and a coast, but the Basque territories are just landlocked specks." Not very good geography, perhaps, insofar as the Basques are concerned, but indicative, possibly, of the perceived limits of "reasonable" secessionism.

One point is clear: a majority of the world's states face devolutionary forces of one kind of another, even as most states are involved in various efforts at multinational union and coordination. Not even the United States is immune: when *Good Morning America* sent me to Hawai'i to observe the January 1993 commemoration of the 1893 annexation, I was amazed at the ferocity of Hawai'ian demands for sovereignty. Some native Hawai'ians want to regain the lands once owned by the monarchy, there to establish a republic—outside the United States.

Devolutionary forces are at work in New Zealand; Australia (where not only Aboriginal rights are at issue but also the matter of continued recognition of the British Crown), is dividing the country and voices for secession are heard, notable in Western Australia); in the Philippines; in Brazil (where some dream of a separate country consisting of the three Southernmost States); and in dozens of other countries. What all these cases have in common is this: dissatisfaction with, and lack of confidence in, the state. It is a global phenomenon, and given the current search for a New World Order, it could not have come at a worse time.

STATE OF IRRELEVANCE?

Some of my colleagues in political geography predict that the current number of independent states (190, according to the Department of State) will approximately double over the next 25 years. Larger numbers of smaller states, they anticipate, will join in new international associations to further their common interests, much as the OPEC states tried to do. In a quarter of a century, they anticipate, there will no longer be a Belgium, an Italy, a Canada, a Nigeria, a Sudan, a South Africa, or a Malaysia—and many other familiar names will either disappear from the map or be taken by remnants of larger states. (Note that the Serbs of Belgrade persist in calling their rump country Yugoslavia, the name of the state they helped destroy.)

If this happens, many more states of the world will become "nation"-states: their boundaries will represent ethnic-cultural limits. Already, Slovenia, Slovakia, Latvia, Armenia, Eritrea, and other newly independent states possess this quality, although all contain minorities. Others in the making may include a Tamil state in northern and eastern Sri Lanka; Turkish and Greek states on Cyprus; an independent Flanders; a separatist Zanzibar; a fragmented Angola; a French and Francophone Quebec; and a Zulu-dominated Natal. Indeed, the dissolution of the Soviet empire may continue within Russia itself.

In such a world of regional and local nationalisms, the European model will fade even further from view. Having accomplished secession, smaller nations (now the term becomes appropriate) must join in international blocs, associations, and unions to further interests they could not promote by themselves. In doing so, they will have to do what the European nation-states of the Economic Community are finding it so difficult to do: yield some sovereignty in exchange for common progress and security. Thus the old nation-state form will lose much of its relevance. Interrelationships and interconnections will matter far more. Decisions made outside the state will affect the people far more than many decisions made in the national capital.

Add to this the impact of the technological and communications revolutions, and the state as we have known it diminishes still further. Soon, 300-channel television will bring global exposure to ideas and goods, a revolution comparable to that wrought by the transistor radio—but affecting a world population four times as large. National cultures will be swept up in a wave of globalization. Multinational corporations, some quite possibly protected by their own security apparatus, international and interactive media, and supranational governmental "communities" will become the relevant structures even as the state becomes less germane.

One of the most obvious characteristics of the current world political map is its diversity: there are states as large as Russia, Canada, China, the United States, Brazil, and Australia, and others as small as Luxembourg and Liechtenstein. But listen to the citizens of states, whatever their size, and you will hear complaints: loyalty, patriotism, and allegiance to flag and country are declining virtually everywhere. Support for smaller constituencies, or for social or religious movements, is on the rise. Barely more than 30 years ago, President John F. Kennedy could urge people not to ask what their country could do for them, and recruits joined his Peace Corps in droves to ask what they could do for their country. Today, an American president asking for sacrifice can expect no such reaction. Esteem for political leadership is low; people who can afford it wall themselves off from crime and protect themselves in other ways, because they have lost faith in the state's capacity to ensure a secure and healthy future.

The map of the twenty-first century, therefore, is likely to be far more complicated than the political mosaic existing today. What remains of the European nation-state model will be fractured into many more pieces, but the pieces will be tied together into regional blocs, wider realms, and global networks. And cartographers will no longer have the luxury of delimiting states with a simple, symbolic line.

CHAPTER 11

RUSSIA'S FAILING FEDERATION

When the Soviet Union disintegrated and 15 new names (including Russia's own) appeared or reappeared on the world political map, many observers thought they were witnessing the collapse of the last great empire. They were wrong on two counts. First, Russia itself, even after the loss of its 14 quasicolonial republics, is still an empire, a vast realm dominated by Russians but also home to millions of non-Russians. Second, another great empire is still functioning today—China.

Even after the loss of its vassal republics, Russia remains territorially the largest state in the world, by far. From Sakhalin Island in the east to St. Petersburg in the west, Russia spans 11 time zones. Now *there's* a challenge for any morning television show—*Good Morning Russia* from Vladivostok would be a half day old by the time it was seen in the west!

The great majority of all Russians live in the western part of the country, west of the Appalachian-like Ural Mountains that stretch from the Arctic Sea almost straight southward to the border with Kazakhstan. True, there's a ribbon of settlement along the southern Siberian margin east of the Urals, extending discontinuously all the way to Vladivostok. Even though that ribbon includes some sizeable cities, such as Novosibirsk, Omsk, and Khabarovsk, it doesn't begin to compare to the heartland of the west, centered on Moscow.

Here in the west lay the centers of Soviet power and foci of the empire. Before the Soviet Union broke apart, its total area was more than 8.6 million square miles (nearly 23 million square kilometers) and its population approached 300 million. Now Russia retains 6,593,000 square miles (about 17,000,000 square kilometers) and has just over 150 million inhabitants. Russia remains the world's largest country territorially, nearly twice the size of the United States, but in terms of total population, it has slipped from third in the world (after China and India) to sixth, behind the United States, Indonesia, and Brazil.

As befits a country so large, Russia has numerous neighbors, from Norway in the west to North Korea in the east. Among these neighbors are two of the giants of the former empire: Ukraine, with a population of more than 50 million, and Kazakhstan, with a territory of more than 1 million square miles (2.6 million square kilometers). Each possesses a residual nuclear arsenal. Both these neighbors are of urgent concern to Moscow, because each contains a large population of Russians now under non-Russian rule.

INSIDE RUSSIA TODAY

When Russia was kingpin of the Soviet Union, its official name was Russian Soviet Federative Socialist Republic (for obvious reasons the acronym RSFSR was the preferred version). The word federative seemed incongruous. Wasn't the former Soviet Union in effect a centralized, dictatorially ruled state? What federal rights did the constituent parts really have?

On paper, they had substantial rights, including the right of secession. If you look back at the old map of the Soviet Union, you will see that every one of the 14 non-Russian republics had a window, either on international waters or on the border of a neighboring, non-Soviet country. None was completely surrounded by other Soviet territory. In practice, of course, secession was unthinkable. Moscow's despots ruled absolutely and rigidly, and whatever the architects of the USSR in the 1920s had in mind, the Soviet Union was a monolith.

In truth, the Soviet Union was a federal state in name only, and it was a democracy only under communist definition of that term. Self-deception was a hallmark of Soviet society, and it started at the top. But the consequences of this extend far beyond the end of the Soviet system: the Russians have no experience in federal organiza-

THE SOVIET UNION: 1924–1991

For 67 years Russia was the cornerstone of the Soviet Union, the Union of Soviet Socialist Republics (USSR). The Soviet Union was the product of the Revolution of 1917, when more than a decade of rebellion against the rule of Nicholas II led to the czar's abdication. Russian revolutionary groups were called *soviets* (councils), and they had been active since the first workers' uprising in 1905. In that crucial year, thousands of Russian workers marched on the czar's palace in St. Petersburg in protest, and soldiers opened fire on them. Hundreds were killed and wounded. Russia descended into chaos.

The czar's abdication in 1917 was forced by a coalition of military and professional men. Russia was then ruled briefly by a provisional government. In November 1917, the country held its first democratic election ever—and, as it turned out, the last for more than 70 years.

The provisional government allowed the return to Russia of exiled activists in the Bolshevik camp (there were divisions among the revolutionaries): Lenin from Switzerland, Trotsky from New York, and Stalin from internal exile in Siberia. In the political struggle that ensued, Lenin's Bolsheviks gained control over the revolutionary soviets, and this ushered in the era of communism. In 1924, the new communist empire was formally named the Union of Soviet Socialist Republics, or Soviet Union in short.

Lenin, the organizer, who died in 1924, was succeeded by Stalin, the tyrant, and many of the peoples under Moscow's control suffered unimaginably. In pursuit of communist reconstruction, Stalin and his elite starved millions of Ukrainian peasants to death, forcibly relocated entire ethnic groups, and exterminated uncooperative or disloyal peoples. The full extent of these horrors will never be known. Many of the country's most creative people were executed.

On December 25, 1991, the inevitable occurred: the Soviet Union ceased to exist, its economy a shambles, its political system shattered, the communist experiment a failure. The last Soviet President, Mikhail Gorbachev, resigned, and the Soviet hammer and sickle flying on the flagstaff atop the Kremlin was lowered for the last time, to be immediately replaced by the Russian tricolor. Eleven former Soviet republics proclaimed the formation of a new union, the Commonwealth of Independent States (CIS), to be headquartered at Mensk (Minsk) in Belarus. Political geographers reasoned that weak centripetal and strong centrifugal forces would soon erode the CIS into ineffectiveness.

tion or operation. Yet federalism may be Russia's only hope as it formulates its future as a multinational, multicultural state—barring a return to authoritarianism.

Here's the problem. The new Russia is a patchwork of internal units, virtually all of them leftovers from the Soviet era, and the relationships between many of these units and the center in Moscow is, to say the least, uneasy. When the communist planners designed the USSR, they not only created such Soviet Socialist Republics as

Armenia, Georgia, and Belarus outside Russia's boundaries, they also established dozens of Autonomous Soviet Socialist Republics, Autonomous Regions, and other entities inside Russian borders. As time went on, the Soviets tinkered with the system, changing boundaries, elevating or lowering the status of republics or regions—sometimes as a reward for loyalty or productivity, sometimes to punish, for example, for alleged collaboration with the Germans during the Second World War. When the Soviet system came to its inglorious end, Russia was saddled with this complex hierarchy of internal entities, numbering, by latest count, nearly 90.

WHERE THE REPUBLICS ARE

The units at the top of this burdensome hierarchy, the republics, occupy some critical space on the present-day Russian map. A cluster of six republics lies east of Moscow, with Tatarstan at their center. Another cluster lies in the Caucasus, extending from the shores of the Caspian Sea almost to the Black Sea. Still another group adjoins Russia's southern border with Kazakhstan and Mongolia. In the north, Karelia abuts Finland, and Sakha, with an area about as large as Kazakhstan, covers most of eastern Siberia, from the shores of the Arctic Sea to within 200 miles (320 kilometers) of the Chinese border.

In 1993, two additional territories proclaimed themselves republics, petitioning parliament—just before President Boris Yeltsin dissolved it—for recognition. One of these is the Urals Republic, centered on the city where the last czar and his family were executed, Yekaterinburg. More ominously, the other self-proclaimed republic lies in the Russian Far East, namely the Maritime Republic, with its capital at Vladivostok, Russia's principal Pacific port and the terminus of the great Siberian Railroad.

During the Soviet era, autonomous republics and regions were established in part to acknowledge the presence of minorities. The Tatar Republic (Tatarstan), for example, was created for the Tatars living in and around the old Muslim capital of Kazan as a reward for joining the revolution against the Czars. The Tatars are descendants of the Mongol Golden Horde who formed the so-called Kazan Khanate, a Muslim state in the middle Volga Valley, in the thirteenth century. In 1552, Czar Ivan the Terrible overran the khanate, destroying about 400 mosques, some of them architectural treasures. Afterward the Tatars sporadically rose in revolt against the czars, but were always defeated—until the communist revolution, when they were on the winning side. This led to the establishment of their republic, smaller than the original khanate but, the Tatars knew, better than nothing. Before long, however, Soviet-communist policies began to afflict the Tatars. Moscow's official atheism ran counter to the Tatars' Muslim revival. Industrial expansion, large-scale agriculture, and oil exploitation brought hundreds of thousands of Russians to Tatar soil. By the time the Soviet system collapsed, the Tatars were outnumbered by Russians in their own republic (48 to 44 percent of the total of just under 4 million). And anger about cultural Russification was widespread. When the external republics (Estonia, Ukraine, and Georgia) were given independence, the president of the Tatar republic demanded the right to equal status. And in 1992, when the Russian Federation Treaty was drawn up to create the present "federal" state, the Tatars refused to sign it

The Tatar Republic lies at the heart of an important zone of industry, mining, oil production, and farming. Its neighbor to the east, the Bashkort Republic (Bashkortostan), lies at the focus of Russia's critical oil pipeline network. There, however, Tatars and Bashkorts (a Turkic people who have lived in this area for centuries) outnumber Russians, and nationalism is on the rise.

The republics along the Caucasus in the south are important —and troublesome—for other reasons. Territories and populations tend to be smaller here, but anti-Russian sentiment is strong and internal divisions and rivalries are powerful. In 1991, Russian forces engaged in what may come to be remembered as the first

TRANSCAUCASIA

Between the Black Sea to the west and the Caspian Sea to the east lies a geographic region of enormous physiographic, historical, and cultural complexity. Physiographically, this region is dominated by the rugged terrain of the Caucasus Mountains, in whose valleys countless struggles have been waged. Historically, this has been a battleground for Christians and Muslims, Russians and Turks, Armenians and Persians. Culturally, the region is a jigsaw of languages, religions, and traditions. It is, by many measures, the Balkans of Asia.

The geographic name for this turbulent region is *Transcaucasia*, signifying its situation astride the great ranges that define its physiography. Transcaucasia at present consists of three political entities: Georgia, Armenia, and Azerbaijan, all former Soviet Socialist Republics in the communist empire.

In the mid-1990s, Georgia, the country that faces the Black Sea, has proclaimed its "Europeanness," and its embattled government has stated its intention of seeking closer ties to Europe. Armenia, landlocked and fragmented, is a Christian country nearly surrounded by Muslim domains. And Azerbaijan is actually the northern half of a larger province, the southern part of which lies in Iran today. So pressing and numerous are the unresolved territorial issues in Transcaucasia that its future is fraught with danger.

wave of "ethnic cleansing" in the tiny Ingush Republic, on the border with Georgia, when hundreds of Muslim villages were destroyed and the population driven into the mountains, where many perished. In the meantime, those who oppose Russian control have formed the Muslim Confederation of Caucasian Mountain Peoples. Remoteness, difficult terrain, and strong cultural forces will make it difficult for Moscow to integrate the Caucasian republics in any Russian federation.

The problems Moscow faces are exemplified in Chechnya, which until 1992 was part of the so-called Chechen-Ingush Republic. Old Muslim enemies, the Chechens and the Ingush heroically stood together and repelled the Cossacks fighting for the czars. By 1870, however, the Russians prevailed. The two ethnic groups were forcibly united by the Soviet-communist planners in 1936. In 1944, Stalin accused the Chechens of collaborating with the Nazis and ordered the entire nation, then numbering about 700,000, exiled to Central Asia. Packed into railroad cars, an estimated 240,000 died en route or shortly after arrival in Kazakhstan.

During Krushchev's regime, the Chechens were allowed to

REPUBLICS IN SOUTHWESTERN RUSSIA

—— Road

⊢⊢⊢ Canal

National capitals are underlined

0   50   100   150   200 Kilometers

0   50   100 Miles

return to their homeland, and in 1957 they and their Ingush neighbors were awarded a joint Autonomous Republic. Chechen nationalism had not faded, and in 1991, soon after the Soviet collapse, Chechen separatists moved to install their leader as president of the Republic. President Yeltsin thereupon signed a decree imposing a state of emergency and voiding the Chechen president's appointment. Before long the Ingush, who also were not enamored of the notion of life in a Chechen-dominated country, took up arms against their old adversaries. Moscow's solution was to split the Chechen-Ingush Republic into two, Chechnya and Ingushetia.

In this breakup, the Chechens got the better deal. Their republic included the capital, Groznyy (450,000), one of the Russian Federation's major oil-refining centers, and some major gas reserves. But the Chechens were not satisfied. They refused to join the Russian Federation, and their aggressively nationalist leader, General Dzhokhar Dudayev, vowed never to allow his country to be taken over by Russian "outsiders" or by Russian "puppets" within his republic. An estimated 220,000 Russians remained in Chechnya even after Dudayev's defiant stand began.

Bristling with arms, filled with gangsters who have links with organized crime in Russia proper, flush with money, and fired by fervent nationalism, Chechnya became Russia's "mouse that roared." After trying to influence the course of events by supporting local opposition to Dudayev financially and logistically, Moscow in December 1994 decided to intervene militarily. A large Russian force moved into the republic, and warplanes attacked Groznyy and other targets. Enormous damage was caused, and hundreds—perhaps thousands—of civilians were killed.

In Moscow, meanwhile, the Russian assault on Chechnya drew heavy criticism. Memories of Afghanistan lingered, but this was a war with a difference: television and the press provided vivid images of what was happening on the flanks of the Caucasus. Some Russian officers refused to move on Chechen villages, and were filmed hugging locals in front of their stopped tanks. Russia was in turmoil; other republics, including some in the Caucasus nearby, watched Chechnya's struggle restively.

Chechnya presented the Russian government with a crucial issue: could the federation survive if its components had to be forced into the fold by force of arms? But if Moscow's campaign in Chechnya failed, how could the federation survive other efforts at secession? Russia's Soviet legacy is no simple matter.

No matter where the republics are, they present Moscow with current and latent problems. In Karelia and in the Mongolian borderlands, boundary issues loom. In Sakha (formerly known as Yakutia for the minority Yakuts whose home this is), resources are plentiful, the Russian presence is weak, and locals demand the right to negotiate directly with interested foreigners. In the pending Maritime Republic, people are fed up with Moscow's strictures; they want a share of the action on the Pacific Rim.

MANIPULATING THE SYSTEM

When Russia lost its colonial empire and Moscow began to restructure the state, there were 16 republics within its borders, five Autonomous Regions, 10 Autonomous Areas, 49 other regions, and six territories. President Yeltsin approved the elevation of the five Autonomous Regions to republic status, creating the 21 now recognized—but he faced demands for recognition not only from the self-proclaimed Urals and Maritime Republics but also from the peoples and leaders of several Autonomous Areas, who felt themselves slighted. While the president conducted his difficult negotiations with the Russian Parliament, he also had to consult with the leaders of the nearly 90 entities within the state. When Yeltsin dissolved the parliament, leading to the brief but deadly revolt of October 1993, he also made it known that he would rein in the powers of the regional leaders.

But the 1992 Russian Federation Treaty establishes the rights of these internal entities quite clearly, and the question has become one of trust: will President Yeltsin and his successors continue to recognize these rights? Or will future Russian rulers and parliaments erode them away?

The republics and regions have much to lose. During the political standoff that ended so violently in Moscow, the republics in some

SOVIET ADVENTURE

My first visit to the Soviet Union was in the summer of 1964. Like many of my colleagues, I had doubts about the fairness of reportage in American media about the USSR, and I wanted to see the situation for myself. Also, still shocked by the assassination of President Kennedy the previous November, I wanted to see where Lee Harvey Oswald had lived, perhaps to learn something about his background or motive.

I got more than I bargained for. The entry from Helsinki, Finland, by bus to Leningrad produced a four-hour inspection and interrogation at the forest-cleared, fenced, barbed-wire-girdled border. My accommodations in Leningrad were the filthiest I had ever seen, the communal bathroom unusable. At the university, my rank as Assistant Professor was insufficient to secure an appointment with the Professor in the Geography Department I had hoped to meet, the Africanist A. Davidov, although I had corresponded to prepare for the visit. "You may wish to meet with Lecturer Yavlinsky," said the helpful secretary. Later, at the University of Moscow, I met with African students studying there. They told me of the rigorous segregation of ranks, a new experience for them, too. The Workers' Paradise was no scholars' heaven.

In Mensk I learned that curiosity could be dangerous. On the basis of my limited information, I got to what I thought might have been Oswald's street; there I made the mistake of explaining to someone who asked in English what I was doing. A crowd formed, I could not leave, and the police arrived to make it an extremely unpleasant day.

I arrived at my Intourist-approved hotel in Kiev to find that no one would register me, speak to me, or provide any service at all; my luggage was removed and not returned. After a night in the uncomfortable lobby lounge I began to understand: the Gulf of Tonkin incident had taken place in Vietnam and the staff awaited instructions on how to deal with foreigners. After another day's wait, things suddenly normalized.

By night train to Zaporozhye, from there by bus to Yalta, and thence by air to Yerevan. Everywhere lay the toppled statues of Stalin: Krushchev's campaign against the tyrant required their destruction. I took a rickety bus along the Georgian Military Highway across the Caucasus from Tbilisi, Georgia, to Pyatigorsk, one of the scariest rides ever but providing a glimpse of towering Mount Elbrus; and then it was on to Rostov, Volgograd (recently renamed from Stalingrad), Kharkov, and back to Moscow.

I came back shaken. If anything, I felt, the U.S. media were too kind to the Soviets; eventual war, I believed, was inevitable. There were microphones in hotel rooms (how could the state recruit the thousands who must be listening?); there was no freedom of expression; self-deception and fear seemed to be endemic.

But there were wonderful moments, too. In Moscow's GUM department store I bought an excellent violin and bow, rosin, and extra strings for less than 20 rubles ($30 U.S. in those days). I had two suitcases, so I put the fiddle in a plastic laundry bag and carried it looped around my neck. It led to countless conversations, many by sign language, and much armwaving, and everyone was eager to teach me Russian tunes. On the train back from Kharkov, I had a whole carful of people singing the night away.

But overall, the atmosphere was oppressive, and images of America were, it seemed, far more skewed than were American impressions of the Soviet Union. I saw no chance for rapprochement, and to me the events of the late 1980s remain nothing less than a miracle.

cases had begun to take on the trappings of independent states. To take the example of Tatarstan again: if you were to fly from Moscow to the capital, Kazan, you would board an Air Tatarstan flight, hear the flight attendant read safety instructions in Tatar, then Russian, see the Tatar flag, not the Russian flag, flying over the airport tower. In the republic, Tatars made huge economic gains, taking businesses over from Russians and securing the right to sell part of its oil direct, rather than via Russia. The Tatar president has visited Turkey as a head of state, reviving the Muslim link and seeking direct economic ties. Elsewhere, too, republic leaders became high-profile personalities promoting the interests of their peoples and reminding Moscow of the many wrongs committed against local peoples during communist times.

If the new Russian constitution, formulated after Yeltsin's dissolution of the parliament, is to give substance to Russia's claim to be a federation, many of the rights to which the republics and the regions have become accustomed will have to be protected. But Yeltsin's pronouncements after that event have been Gorbachev-like in their conservatism, and he may leave Russia a legacy of centralism that may eventually pave the way for another era of authoritarianism. Once again, Russia would be a federation in name only.

A NEW RUSSIAN EMPIRE?

Russia's total population is listed as just over 150 million, ranking the country, as noted, sixth in the world. But nearly 30 million Russian citizens are not ethnic Russians: they are minority nations and groups caught up in the web of past Russian imperialism. Today, cultural geographers count nearly 100 such minorities in the country, including 40 large enough to claim the status of nation. And several of these minority nations have historic and cultural links to peoples outside Russia.

Throughout Russia, religious, linguistic, and tradition revivals are under way. Even as the Russian Orthodox churches are reborn, reclaiming properties they once owned and renovating churches that survived communism, other faiths elsewhere are astir. The president of Kalmykia, a republic that adjoins the Volga River and the Caspian Sea, touts his country as "the only Buddhist state in Europe." His people, who make up about 43 percent of the population of under a half million in this South Carolina–size republic, have historic links to Tibetan Buddhists. From the Finns in Karelia to the Mongols in Buryatia, Russia's hinterlands form a mosaic of nationalities and cultures.

Being widely dispersed geographically, the capacities, aspirations, and traditions of these republics and regions vary greatly. Many of the nationalities were ruthlessly suppressed not only by the czars who conquered them, but also by the communists who ruled them for seven decades of the twentieth century. Stalin, for example, also exiled the entire Kalmyk nation to Central Asia in retaliation for their alleged support of the German invaders during the Second World War; tens of thousands died. Some republics acquired a considerable industrial base during Soviet times; others were developed agriculturally; mines were opened in many. All of them, to a greater or lesser degree, were occupied by Russians, who tended to control the political system and the economic apparatus.

Much of this could be said about the external republics, the ones outside Russia that were part of the Soviet Union, as well. And that's the worry: ultimately, the Russian view of the provinces will be an imperial, not a federal one. The former external Soviet republics could be cast adrift when the USSR collapsed because they lay outside Mother Russia geographically. The republics and regions that remain lie within Russia, and while several lie on external borders or coasts, others are landlocked. Secession is unthinkable, but some autonomy in a federal framework is the hope.

The struggle between federalists and centralists, which is likely to last for years, is the defining contest for the future of Russia. To the question of whether federalism will prevail in a state that calls itself a federation, there is as yet no answer. Again, visions of the end of empire are premature.

RUSSIA'S NEAR-ABROAD

"Where Does Russia end?" asked The *Economist* in its issue dated November 13, 1993. The journal could not come up with an answer, and for good reason: the Russian presence and influence in the old Soviet republics outside its borders continue. The Russians call this neighboring tier of countries the near-abroad, and Moscow has made it clear that it regards this area as its sphere of influence.

President Gorbachev started, and Yeltsin continued, an effort to link the 14 former Soviet republics with the fifteenth, Russia, in the Commonwealth of Independent States (CIS), to be headquartered, initially, in Mensk (Minsk), capital of Belarus. Although signatures were gathered and meetings held, and some sports teams were sent to compete internationally under the CIS banner, the organization attained little substance. The Commonwealth is no European Union—which is understandable. The EU took a half century to reach its present, still-incomplete form. The CIS was formulated almost overnight.

More consequential is the presence of nearly 30 million Russians in these former colonies and a growing concern over their fate. Take the case of Ukraine: when this former Soviet republic became an independent state in late 1991, it took out of the empire a huge industrial complex, vast, fertile farmlands, and over 50 million people—more than 10 million of them Russians. Importantly, this Russian population is clustered on the Crimean Peninsula and adjacent areas. How did these Russians come to find themselves in Ukraine? The main reason: until 1954, the Crimea was Russian territory. But in that year, Premier Nikita Krushchev decided to reward Ukraine by giving it the Crimean Peninsula and redrawing the boundary accordingly. After all, this was just an exchange of land among sister republics of the Soviet Union; everyone would remain under Moscow's control. In his wildest dreams, Krushchev could not have imagined that his gesture would exile millions of Russians to foreign residency.

Today, the political-geographical situation caused by Krushchev's impetuous act is fraught with danger, but Ukraine suffers from other political fault lines as well. The name of the country is derived from the Russian word for margin or edge; the best geographic translation is "frontier." And Ukraine is just that: a land of transitions, an artificial creation in some ways reminiscent of another—Yugoslavia. Welded together are pieces of former Polish territory in the west and Russian territory in the east; incorporated in the non-nation are Belarussians in the north, Hungarians and Poles in the west, Moldovans and Romanians in the south. A piece of Slovakia was cut off in 1945 and attached to Ukraine, and that area, Transcarpathia, formerly Ruthenia, remains a tapestry of minorities. In Ukraine, unlike nearby Georgia, no minorities were given republic status, so the associated dangers seem more remote. But the map of Ukraine hardly inspires confidence. The brief surge of Ukrainian nationalism that led to independence in 1991 was centered in the west, where the city of Lviv (Lvov) proclaims itself the birthplace of the nation. There was much less enthusiasm for it in the south and east, from the environs of the Black Sea port of Odesa (Odessa) through the Crimea Peninsula to the industrial Donbas. In the north-central capital Kyyiv (Kiev), ambivalence seems to rule these days—ambivalence about economic reform, about ideology, about the Crimea question, even, it seems, about independence.

Might Moscow be emboldened to take advantage of Ukraine's weaknesses? It would not be the first time. Ukrainian nationalists remember 1918, when a newly independent Ukraine was seized by Russia while the Western powers stood by. Ukrainians also remember the late 1920s and 1930s, when Stalin doomed the country's peasantry through an imposed famine that killed millions. They also recall the 1940s, when the Tatars still living in the Crimea Peninsula were accused of collaboration with the Nazis and ruthlessly deported to Central Asia. The Crimean Tatars suffered the same fate as the Kalmyks. Ukrainians also know that the massive industrialization of Ukraine's eastern flank and the militarization of the Crimean ports have created not only a major Russian presence but also a dependence that severely limits Kyyiv's freedom of action. For example, Ukraine needs Russia's energy resources, its oil and natural gas; during the Soviet era, a pattern of supply and consumption was established that endows the supplier with a powerful weapon.

No other country in Russia's near-abroad comes close to Ukraine in terms of importance or conflict potential, not even Kazakhstan. Militant nationalism is on the rise among Russians in the Crimea Peninsula, and this is infecting other Russians elsewhere in the republic. The disposition of Russia's Crimea-based fleet and Ukraine's nuclear weapons remains at issue. Tatars and their descendants are returning to the Crimea to demand the return of land now owned by Russians. Meanwhile, the Ukrainian government has presided over an economic collapse far deeper than that afflicting Russia. Again, as so many times in the past, Ukraine's weaknesses invite external intervention, and again Ukraine seems to lack the powerful friends it needs to survive. In one bold stroke, Moscow could recover a piece of an empire in this vacuum on its periphery. Russia's actions toward this former colony will tell the world what kind of order will prevail.

A look at the map suggests that there is only one other country in Russia's near-abroad capable of testing Moscow's constraint to a similar degree: Kazakhstan. Ukraine is territorially large; viewed as part of Europe, it is its largest state, and its population of 52 million ranks among Europe's biggest as well, but Kazakhstan is more than four times as large. And while Ukraine's population is about 20 percent ethnic Russian, Kazakhstan's is nearly 40 percent Russian. While Kazakhstan is a territorial giant (it covers an area about one-third the size of the United States), its population is just approaching 20 mil-

TURKESTAN

POPULATION

- Under 50,000
- 50,000–250,000
- 250,000–1,000,000
- 1,000,000–5,000,000
- Over 5,000,000

National capitals are underlined

0 150 300 450 600 750 900 Kilometers
0 100 200 300 400 500 Miles

lion. It is the geography of this population, though, that gives rise to concern over the future. The great majority of Russians in Kazakhstan live in the northern one-third of the country, where it adjoins Russia. Russian numbers are just about matched by Muslim Kazakhs, who dominate in the vast southern reaches of the country.

As the above map shows, the north of Kazakhstan has been

Russified to the point of virtual integration with the Russian state. Major transport routes between western and eastern Russia stretch across their zone; cities and towns have Russian names and Russian cultural landscapes. The Soviet space program was centered in Kazakhstan, and Russia has continued to use the facilities.

How long can a representative government, based in the remote Southern capital of Almaty, accommodate this situation? The capital is being moved westward. Kazakhstan, like Ukraine, is itself a frontier: here a revived Islam meets the reborn Russian Orthodox Church; here Russians and Kazakhs share the land with as many as 100 smaller minorities; here lie the scars of decades of Soviet agricultural mismanagement. When Russian and non-Russian interests diverge, as they will, another crisis will threaten the near-abroad.

Can trouble be averted? One potential is geographic: the relocation of the Russia-Kazakhstan border. Feasibility studies to that effect should be started now, including compensation prospects for the Kazakh south. Boundary-making after hostilities break out is difficult, as Bosnia has reminded us.

Other Russian minorities in the near-abroad are smaller, but face uncertain futures, too. From Estonia to Kyrgystan, the Russians were colonists who represented Moscow, communism, imperialism, atheism, and oft-unwanted change. And in many places, they now confront the kind of resentment faced by Europeans at the time of decolonization. In Estonia, more than one-quarter of the population is Russian, strongly concentrated in the country's east; the Estonian government has challenged this Russian population to learn and use the Estonian language as a precondition for citizenship. In neighboring Latvia, where the Russian population constitutes about one-third of the total and is widely dispersed throughout the country, relations between Latvians and ethnic Russians are also strained. In Moldova, local Russians (13 percent) tried to create their own, internal republic on a narrow strip of land between the Dniester River and the

KALININGRAD

The Russian exclave of Kaliningrad lies wedged between Lithuania and Poland, facing the Baltic Sea through its gigantic naval port. Soon after the Soviets acquired this German base in 1945, virtually all ethnic Germans were expelled. Not much was left of their cultural landscape: relentless British bombing had devastated the place.

Russians replaced the departed Germans, and today Kaliningrad's population approaches 1 million people, 90 percent of whom are Russian. In the Soviet political scheme of things, Kaliningrad had the status of an *oblast* (province), and it was made part of the Russian Republic—as it remains today.

Uncertainty now prevails in Kaliningrad. Independence is not an issue, but that could change. Despite the destruction of much of seven centuries of German heritage, the German past still can be felt: the renowned philosopher, Immanuel Kant, lived and worked here and lies buried in the ruins of the German cathedral. The city's German name, Konigsberg, is used frequently.

In 1990, Moscow decided to resettle some 20,000 ethnic Germans from Kazakhstan to Kaliningrad. Some political leaders in the oblast worry that the German influence in their territory will threaten the Russian majority. There is also talk that Kaliningrad might become a Russian free-trade zone, revitalizing the commercial port. Could Kaliningrad become the Hong Kong of the Baltic?

Ukrainian border—the Trans-Dniester Republic—to avoid reprisal from Romanian Moldovans.

The most isolated Russian minorities, though, are in former Soviet Central Asia. In Tajikistan, on the border with Afghanistan, the Russian army has already intervened on behalf of the old guard, still ensconced in the republic. Elsewhere, the Russian minorities are diminishing remnants of the colonial era, starkly different from the reviving cultures of Muslim Turkestan. For these people there is concern in Moscow, especially in view of the rise of Islamic fundamentalism here.

As a result, the Russian military wants to maintain a presence in the near-abroad, preferably by means of a string of Guantanamo-like military bases. This runs counter to the wishes of the republics, but they may not have a choice. Already, the Russian military has followed its intervention in Tajikistan with decisive intervention in Georgia—not to save Russians from retaliation, but to save the Georgian government headed by former Soviet Foreign Minister Eduard Shevardnadze.

The course of events in Georgia is an ominous sign of what may lie ahead in the near-abroad, although the issue here was not the fate of a Russian minority. During the Soviet era, Georgia was one of the 15 Soviet Socialist Republics, and to acknowledge the status of minorities within Georgia, two so-called Autonomous Soviet Socialist Republics were created and one Autonomous Region. Soon after independence from Soviet rule, anti-Georgian violence erupted in the South Ossetian Autonomous Region, a landlocked area in the north-central border area. But worse was to come. While an armed conflict between political factions took place in and around the capital of Tbilisi, the internal Autonomous Republic of Abkhazia, on the Black Sea coast in the northwestern corner of the country, declared its independence and its secession from Georgia.

Abkhazia is home to one of the many rather small minorities that make up the human geography of Transcaucasia—the land between the Black and the Caspian Seas. History, geography, and a window of opportunity emboldened the Abkhazians to seize control of their territory. Georgia's military was divided and weakened and without strong leadership. But the key moment came in September 1993, when Moscow was in the throes of the rebellion by conservatives and President Yeltsin was under siege. While all eyes were on the capital, Russian commanders quickly supplied the Abkhazian separatists with weapons; they also allowed mercenaries from the border republics, including North Ossetians, to cross the Russian-Georgian border into Abkhazia to help beat back the Georgians. Then, when the Abkhazian campaign succeeded and the separatists prevailed, a back-to-normal Moscow agreed to help keep the peace in the rest of Georgia—in return for a permanent military presence on Georgian soil. Today, Moscow stations about 5,000 troops in Georgia at five military bases and three of the country's ports. And Georgia was forced into signing the CIS Treaty it had refused to ratify since independence.

While Abkhazian representatives appealed for international recognition of their republic (none was forthcoming), events in Georgia underscored the dangers inherent in the fermenting near-abroad. No Russian minority was threatened there, and Russian military hard-liners appear simply to have acted opportunistically; perhaps some thought that Russian territory might expand at Georgia's expense. But what if the Abkhazian republic had a substantially Russian population? As it is, an important country in Russia's near-abroad has lost not only territory, but also much of its sovereignty. It may be a harbinger of what is to come.

It is yet unclear how much power or influence the Russian military will have in the Russia of the late 1990s and beyond, but the justifications of a kind of Russian Monroe Doctrine in the former Soviet republics already exist. And the lesson of Georgia is unlikely to be lost on the internal republics—on Tatarstan, the aspiring Maritime Republic, and the others. Add the external challenges to the internal problems of this devolving empire, and the prospect of a genuine federal state slowly fades from view.

C H A P T E R   1 2

THE COMING CHALLENGE OF CHINA

During the first half of the twentieth century, the attention of geographers on both sides of the Atlantic was riveted on the next installment of an engrossing debate about the future of Eurasia. It all started one snowy January night in London in 1904, at the Royal Geographical Society, when a Scottish geographer named H. J. (later Sir Halford) Mackinder delivered a lecture entitled "The Geographical Pivot of History." The dreadful winter weather kept Mackinder's audience rather small, but from the minutes of the meeting it is clear that those present were aware that they had heard something extraordinary.

Some months later, Mackinder's address was published in the Geographical Journal, complete with the comments of those who responded and queried the speaker that evening. Its publication evoked a storm of reaction, and for decades "The Geographical Pivot of History" was undoubtedly the most oft-quoted article ever published in this field. Mackinder himself participated, revising his position as time went on, but convinced that he had read the future accurately. The debate continued well after Mackinder's death in 1947, and even today his ideas are cited. As recently as 1981 there appeared a massive volume entitled Geopolitics and War: Mackinder's Philosophy of Power. What Mackinder set in motion is still going strong.

What did Mackinder say to precipitate so much discussion? First, at a time when naval forces had secured seaways for imperial powers and had given Britain preeminence in the world, he predicted that land-based, impregnable strongholds would ultimately dominate global geopolitics. And second, his analysis of the distribution and quantity of known resources around the world led to the conclusion that interior Eurasia (the "pivot area") alone contained the wherewithal for a future power to challenge the world for total domination. This pivot area—he later called it the *Heartland*—lay in Eastern Europe and Western Russia. At a time when Russia was a collapsing, overextended state losing a war against Japan on its eastern flank, Mackinder foresaw the rise of a world power based in its core area.

To achieve such power, Mackinder said, an inner Eurasian state would have to unify and control Eastern Europe; this was the key. And so his argument was summarized in a ditty that still appears in textbooks today:

> *Who rules East Europe commands the Heartland*
> *Who rules the Heartland commands the World-Island*
> *Who rules the World-Island commands the World.*

The world-island was Eurasia. Inevitably, Mackinder reasoned, the seat of a globe-dominating power would be on the Earth's largest landmass.

At first, Mackinder's critics argued that he had miscalculated the resource potential of areas outside his heartland. Later, as the technologies of war changed, his view of strategy was challenged. But Mackinder never swayed from his basic position. And when the Soviet Union emerged after his death as one of the two power cores in a bipolar world, Mackinder's original article got a lot of careful rereading.

THE PACIFIC RIM

It was during this fascinating exchange about power and politics that the notion of a Eurasian Rimland first arose. A military strategist and geographer named N. J. Spykman was, I think, the first to use the word "rim" for the periphery of Eurasia. In a memorable response to Mackinder, Spykman, in a book entitled *The Geography of the Peace* (1944), also waxed poetic:

Who controls the Rimland controls Eurasia
Who rules Eurasia controls the destinies of the World

Spykman, of course, had the advantage of 40 years of further observation. But his, too, was an early alert to changing realities. No matter how powerful the Soviet Union would emerge from the Second World War, Spykman wrote, its strategy should be to dominate the Eurasian Rim, eastward as well as southward. Without power in the Rim, Moscow's dominance would ultimately fail.

The heartland-rimland debate did not remain confined to the scholarly literature. Newspapers and magazines followed the controversy eagerly and printed summaries of the latest salvos. But others with more practical objectives also monitored it. To German geopoliticians, who are known to have analyzed Mackinder's arguments closely, the heartland theory provided the sort of justification for eastward expansionism that would go well with Friedrich Ratzel's notions of the organic state (see page 192). To Russians, Mackinder's focus on Eastern Europe lent urgency to the drive to control its satellites there.

Today, several years after the end of a half century of great-power confrontation, the focus is on Eurasia's rim, not on the heartland. Those who foresaw the potentials and opportunities of the rim, from Western Europe to East Asia, appear vindicated. After all, it was a rimland country, Britain, that forged a global empire, and another rimland state, Japan, that leapt from stagnation to aggression and rapidly acquired a vast colonial realm.

But the focus has shifted in another, crucial way. In the early 1990s, international trade crossing the Pacific Ocean, by value and volume, surpassed that crossing the Atlantic. That momentous statistic, reported in the back pages of the business sections of newspapers, reflects a critical redirection of the fortunes of the world. Europe is in relative decline, while East and Southeast Asia rise. What we might have called the Atlantic Rim is giving way to the Pacific Rim as the hub of economic growth.

Once again, geography, an ally of the United States in so many ways, is proving to be a friend. Our northeastern core area, with its great cities and major industries, is a legacy of the times when people and innovations streamed from Europe to American shores. Now a new age is dawning, but there will be no need to superimpose the trappings of this new age upon the vestiges of the old. The Pacific century-to-come will reorient the United States westward, and its landscapes will transform Pacific-facing shores and hinterlands.

It is likely that we will look back upon the Clinton presidency as the time when America's Pacific realignment began. The United States has long had Pacific interests, of course, and in the ill-fated Vietnam War it tried to control the course of political events. But now the challenge is economic, and Japan has taught us the danger of closed markets and unfavorable trade balances.

Thus, in November 1993, Clinton invited the leaders of 15 East and Southeast Asian countries to convene in Seattle to discuss economic and social prospects. Add still another acronym to the lengthening list we use these days: APEC, the Asia-Pacific Economic Cooperation forum. This is a young and still quite informal organization, consisting of the United States, Canada, Japan, South Korea, China, Hong Kong, Taiwan, Australia, New Zealand, and the six members of ASEAN, the Association of Southeast Asian Nations: the Philippines, Indonesia, Thailand, Malaysia, Brunei, and Singapore. In 1992, APEC members agreed to establish, in Singapore, a secretariat for the organization; that was a significant development for regional economic cooperation on the Pacific Rim of the future.

APEC is geographically interesting not only for what it includes, but also for what it excludes. The country with the longest Pacific coast of all, Russia, is not a member. Nor, as recently as 1994, was Vietnam, a country with 75 million people and enormous potential. Also note that not a single Middle or South American country forms part of APEC, not even Chile, whose economy has been growing at Asian rates and whose major trading partner is Japan.

So the picture is likely to change (APEC will lose a member when China absorbs Hong Kong in 1997). But one thing is certain: the long-term fate of the Pacific Rim, and that of the world, will be determined by the role China will play during the coming century. With 1.2 billion citizens within its borders and millions living abroad, with an economy growing at a rapid rate, and with a vast range of raw materials and resources, China will pose a challenge the likes of which the world has not yet seen.

A CHINESE EMPIRE

China, like the former Soviet Union and the globe-girdling British domain before it, is an empire. It is the product of imperial expansion by the People of Han, the formative dynasty of two millennia ago. It is a vast realm that extends from the shores of the Pacific to the borders of Pakistan, and from the deserts of Inner Asia to the snowcapped Himalayas. Han China, or what the British invaders used to call China Proper, is eastern China, the China of Beijing, Shanghai, and Guangzhou, the China of the Yellow, Chang, and Pearl Rivers. But in all directions from this ancient core of Chinese civilization and power, the Chinese domain extends over peoples who were absorbed and assimilated, or subjugated and dominated: Manchus in the north, Mongols in the northwest, Turkish peoples in the far west, Tibetans in the southwest, Southeast Asians in the south. Among China's minorities are Koreans, Kazakhs, Burmans, Thais, and many others.

True to Leninist principles, the Chinese, like the Russians, set aside territories for their minority peoples, although they did not call them republics. Thus China Proper is flanked by a semicircle of Autonomous Regions (the preferred designation here): the Nei

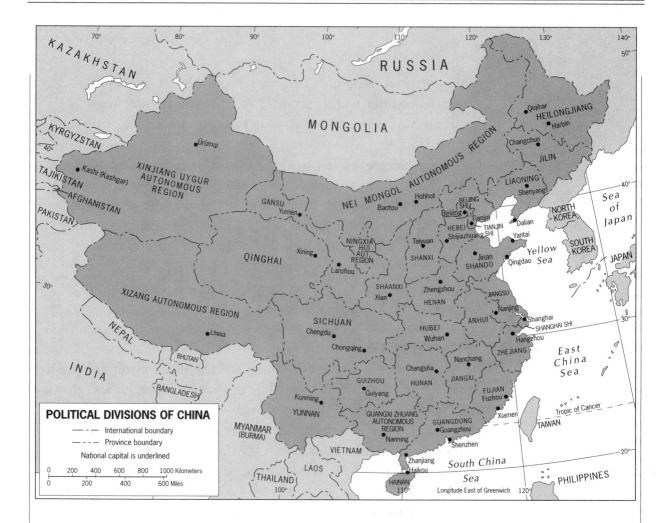

POLITICAL DIVISIONS OF CHINA

— · — · — International boundary

— - — - — Province boundary

National capital is underlined

| 0 | 200 | 400 | 600 | 800 | 1000 Kilometers |
| 0 | 200 | 400 | | 600 Miles | |

Mongol A.R. on the margins of the buffer state of Mongolia, the Xinjiang Uygur A.R. in the far west, adjacent to Turkestan, and the Xizang (Tibet) A.R. are the largest. Smaller A.R.s include the Guangxi Zhuang region adjacent to Vietnam in the south, and the Hainan A.R. off the coast of southern China. In these Autonomous Regions, some of the strictures of Chinese communist dogma were relaxed; for example, the one-child-only policy, enforced with draconian methods in Han China, was not brought to bear here.

Thus the China controlled by Beijing is actually larger than China Proper, and even a casual glance at the map suggests the country's imperial legacy. Moreover, several of China's domestic colonies have been strongly affected by developments on the Pacific Rim. Workers are streaming from the Hainan A.R. toward the island's booming capital, Haikou, which is not part of the A.R. but whose economy is mushrooming. In the coastal province of Guangdong, a similar migration from the neighboring Guanxi Zhuang A.R. is swelling the mass of workers laboring in the new factories. The reverse ripple effect of this is touching the more remote provinces of China proper as well as the Autonomous Regions: these provinces are becoming restive as they see the advantages and opportunities on China's Pacific Rim.

So China, too, faces a transition. Russia's Soviet empire fell apart because of a failing economy and a corrupt political system; the transition envisaged by Gorbachev attracted too few adherents to enable it to succeed. China's rulers are trying a different route, introducing a socialist market economy while retaining Communist Party control over government. What this means is that the Party retains the center's power over Han as well as non-Han China, while permitting (commanding) an economic transformation.

All this is bringing unprecedented change to the economic and social geographies of China. In order to stimulate the economy, the Beijing regime established a number of so-called Special Economic Zones (SEZs), Open Cities, and Open Coastal Areas. Here, investors are offered many incentives. Taxes are low. Import and export regulations are eased. The hiring of labor under contract is allowed. Products made in these economic zones and open areas may be sold on foreign markets and, under certain conditions, in China as well. And profits may be sent back to the investors' home countries. In some of the SEZs, this program has had spectacular results. Of the five SEZs, one, Shenzhen (adjacent to Hong Kong in Guangdong Province), has grown from a fishing village of about 20,000 in 1980 to a metropolis approaching 3 million in 1995, the fastest-growing city anywhere on Earth. And while no other Special Economic Zone or open area can match what has happened in Shenzhen, rapid growth is occurring all along the Pacific Rim of China, from Dalian in the north to Haikou in the south. Skyscrapers tower over the traditional townscapes, streets are jammed with traffic, factories and workshops large and small disgorge streams of new products ranging from toys to computers. A pall of dust and pollution hangs over every city; the noise of growth is, at times, deafening. The Industrial and post-Industrial Revolutions have arrived here all at once.

Overall economic growth in China during the year 1993 was 13 percent, a dizzying rate by world standards. It is straining infrastructures and threatening inflation—but more important, perhaps, it is strongly concentrated in the provinces on the Pacific Rim. There, growth is far in excess of the 13 percent that is the national average; in the interior, it is far less. Within China, the coastal provinces are growing apart from the rest of the country.

When I first visited China, during the summer of 1981, little of this was yet visible. My notes refer to the dilapidated state of Shanghai's port and urban facilities, the low-rise profile of Beijing, the then uncommon sight of Westerners in places like Xian and Chengdu, and the apparently good-natured grousing about the old guard in Beijing by blue-overalled workers at a jade factory at what we still called Canton.

By 1994, my most recent field trip, my impressions were vastly different. Modern high-rises and hotels are creating high-profile downtowns; the old townscapes are being swept away by bulldozers. Across the Huangpu River from Shanghai's waterfront avenue called the Bund, a huge development project is under way, and locals believe that this venture, named Pudong, will someday be a name as famous as Hong Kong is today. (Hong Kong after 1997 will be named Xanggang on the map of China). Two giant bridges already link Pudong to central Shanghai; a huge new coal-fired power plant has been built. A new airport is being laid out, and a system of six-lane roadways anticipates transport needs. Meanwhile, central Shanghai is changing by the day. Skyscrapers tower over the dwindling, tile-roofed traditional buildings of another era. The old British colonial banks and hotels along the Bund have been preserved and are being renovated, and the Bund itself has been transformed into a riverwalk, but nothing else, not even the historic buildings of the era of extra-territoriality, may be safe from demolition. Glass towers rise, often incongruously, over some of the world's most congested streets.

Beijing's transformation is even more astounding. In 1981, the Beijing Hotel, not far from Tienanmen Square, with three architectural imprints (pre-Soviet, Soviet, post-Soviet), was visible from the city's few vantage points. Today, it is dwarfed by a forest of skyscrapers in various stages of completion. In 1981, we spent hours on two-lane roads, competing with handcarts, bicycles, animal-drawn wagons, dilapidated buses, and smog-belching Soviet-era trucks. Today, a network of wide highways shortens travel time significantly. Beijing is taking on the look of a modern capital.

All this development comes at a price. The air over much of China proper is as polluted as any on this planet, and while developed countries fret over environmental deterioration and pass laws to limit it, the Pacific Rim from Dalian to Djakarta is a cesspool of gases and dust. It is not unusual for the sun to be barely visible on a cloudless day, but many of these atmospheric pollutants do more than obscure the sky. They risk causing lasting damage to the global environment.

None of this is likely to worry China's developers. A Chinese colleague, a geographer in Shanghai, told me that the attitude is this: the Western countries used up their share of global pollution rights during the Industrial Revolution in Europe and North America. Now it is China's turn, and when China has reached its own development objectives, there can be talk of pollution controls.

But not until then. In fact, China has joined other countries in the global effort to reduce emissions, (for example through the Rio de Janeiro Earth Summit of 1992, where it was one of the 178 conferees). But in China's runaway development scene, controls of any kind would be hard to impose—even if the Chinese wanted to.

And that is the other side of China's emergence as a global economic and political power. In Canton, now called Guangzhou, the workers and salespeople at that same jade factory still grouse, but now their grousing isn't so good-natured. The bureaucrats in Beijing are corrupt, I was told time and time again, and they're holding up development with their stifling regulations, which seem to be designed to extract money from those able to pay to circumvent them. You can hear this refrain from Haikou to Dalian, and even in Shenzhen—where there is little outward evidence of any stifling of growth.

How serious is this impatience with the central government? The question is whether it reflects a real threat to China's capacity to survive its modernization in one piece. Uneven economic growth happens in all countries, but in China it is far more disproportionate than the norm. Averages for China—GDP, annual rate of growth of the economy, and so on—conceal staggering differences. The southeastern province of Guangdong outproduces all 14 interior provinces combined, for example. The string of Pacific Rim provinces leaves the rest of China in the dust—literally and figuratively. So the communist rulers in Beijing face a daunting task: to guide and control this runaway process to the satisfaction of not only the participants on the Rim, but also those in the interior. There, locals have their own grievances, perceiving favoritism toward the coastal developers. Some may be tempted to resist Beijing's authority. Among my colleagues who specialize in the geography of China, there are those who suggest that a challenge to Beijing's authority may come from a new generation of warlords in China's interior, not from disaffected entrepreneurs on the Pacific Rim. Whatever the future, there is no doubt that the Chinese empire of today is under stress and that the future of China as a unified, centrally governed superstate is far from secure.

A WORLD SUPERPOWER?

If China succeeds where the Soviet Union failed, and if it does survive its transition whole, then the world will confront a superpower of unprecedented strength and dimensions.

Consider China in comparison with previous contenders for world supremacy. Britain in its imperial heyday had, at home and abroad, perhaps 4 percent of the world population. Nazi Germany had greater numbers but less access to raw materials. Soviet Russia had a large population base and a vast array of resources, but it was defeated by the inefficiencies of its political-economic system. And now, in an apparently eastward march of power on the Eurasian landmass, there looms China—a state with a population (counting the overseas Chinese) of nearly one-quarter of the world's, with vast and varied resources yet to be augmented by maritime claims in the South China Sea and further discoveries in the west; and a location at the heart of the Pacific Rim, facing the world's largest ocean. China has the world's largest army, it is a nuclear power, and it is becoming a naval force. And its historic sphere of influence extends far beyond its borders, into Russia, Mongolia, Turkestan, Korea, and Southeast Asia.

But China's Pacific Rim development does not make it a superpower overnight. Annual growth rates are high, but the base from which these rates are measured is low. Average annual income per person in China is still under $1,000 U.S., in the lowest category defined by the World Bank. Also, China's economic growth has moved in fits and starts, booms and busts, as Beijing has tried to keep it under control. And for the present, China depends heavily on foreign investments, foreign manufactures, foreign technology, and foreign ideas. For example, China's booming economy requires efficient long-distance transport, but China's railroad system cannot provide it and China's air routes are under-equipped. Just over a decade ago, there was only one airline in China, using, for internal traffic, mostly aging and unreliable Soviet planes. Today, there are more than 30 airlines, but efficiency has scarcely improved. China will need dozens, possibly hundreds, of large aircraft from the world's plane manufacturers over the coming decade; without these, development will be slowed.

So it is with many of China's other needs. Modernization takes time. On the other hand, it is worth reminding ourselves of the incredibly fast change of Japan from a stagnant, feudal society to a powerful, expansionist state and, in its second transformation in a single century, to a global economic power. In the mid-1860s, Japan did not seem destined to become an Asian power, let alone a global one. Closed to foreign influences, tradition-bound, and internally divided, it had all the ingredients for failure. But then a group of reformers took power, moved the capital to Tokyo, over-hauled the economy, suppressed the long-running rivalries among local rulers, introduced Western technologies, and built a military machine. This was the so-called Meiji Restoration (proving that these reformers also were good politicians, they named their ambi-tious plan after the weak but revered emperor), and it transformed Japan in just a few years. The campaign started in 1868; by 1879, the Japanese claimed their first colony, the Ryukyu Islands, includ-ing Okinawa. In 1895 they took Formosa (Taiwan) from the Chinese. In 1904 they inflicted a disastrous defeat on Russian forces in Sakhalin. And in 1910 they colonized Korea. In a few decades, Japan had risen from a fractious weakling to an imperial power. Soon, Tokyo's forces were in China, making Manchuria another dependency and pushing deep into Southeast Asia. As Americans remember all too well, Japan's aircraft carriers and fighter planes inflicted a devastating blow on the U.S. Navy in Hawai'i in 1941.

Japan's second restoration came after its defeat and destruction in the Second World War, and again it happened with lightning speed. The Japanese transformed their society to compete, but on a different playing field. Once more Japan—a country the size of Montana with a population less than half that of the United States—put itself among world leaders, not, this time, as a rival colonial power, but as an economic juggernaut. True, Japan's economy has been troubled in the 1990s, but it remains the world's second largest, an achievement that should be seen in context of its relatively small size and its extremely limited domestic resource base.

JAPAN AND CHINA

In the mid-1990s, Japan remains the economic giant on Austrasia's Pacific Rim, an industrial powerhouse whose products dominate world markets and whose investments span the globe. Nevertheless, Japan faces problems. Competition from other Pacific Rim countries, including South Korea, Singapore, Taiwan, and Hong Kong, is undercutting Japanese products on world markets, just as Japan's own products once undercut more expensive Western goods. Many of Japan's overseas investments have lost much of their value. And always, Japan's dependence on foreign oil is a potential weakness: rising costs or supply interruptions would have grave consequences for the country's energy-dependent economy.

Nonetheless, Japan still has enormous further potential, much of it arising from its location. The Russian Far East, lying opposite northern Honshu and Hokkaido, is a storehouse of raw materials awaiting exploitation, and the Japanese are ideally located to help achieve this. A major obstacle, however, has been the Kuriles, a group of small islands northeast of Hokkaido (see page 181). Moreover, Japan's technological prowess is finding an ever more sizable market in developing China, and again the Japanese are situated on the doorstep of growing opportunity.

All this should be viewed against the backdrop of a changing social and cultural Japan. The population of 125 million is aging, and demographers report that it will soon stabilize and then begin a slow decline. That is why, as we noted earlier, Japan faces a labor shortage—a situation previously unheard of. And the growing number of elderly people is increasing the costs the state must pay in the form of health care, retirement, and other services for the aged. This also comes at a time when modernization is putting strains on the family. The Japanese have a strong tradition of family cohesion and veneration of its oldest members, but there is much evidence that this custom is breaking down. Younger people want more privacy and self-determination, and they are less willing than their parents and grandparents to accept overcrowded housing and limited comforts. In Japan, for all its material wealth, family homes tend to be small, cramped, often flimsily built, and sometimes without basic amenities considered indispensable in other developed countries. With their high incomes, the Japanese have come to rank among the world's most-traveled tourists, and, having seen how Americans and Europeans of similar income levels live, they have returned dissatisfied.

This dissatisfaction spills over into the workplace and the schools. Japan's work ethic is world-renowned, but it is sustained at a high cost to the individual and often involves enormous demands on time and dedication to the corporate good. In education, the requirements are perhaps the most rigorous anywhere, and the competition is fierce. Now some leaders are beginning to question the appropriateness of such standards and regulations in the new, economic world power Japan.

For the countries and economic clusters along Asia's Pacific Rim, Japan remains the model—a model to be emulated but challenged. Workers in the factories of Taiwan are paid less than workers in Japan, and workers in South Korea are paid less than those in

Taiwan. And so their cheaper products compete with those of the Japanese; but from other places (China, Malaysia) come cheaper goods still. Japan stands at a crossroads of great opportunity on East Asia's threshold and, of growing rivalry from its Pacific Rim neighbors to the south.

THE FOUR TIGERS

Until about two decades ago, Japan's dominance in the industrial geography of the East Asian Pacific Rim was beyond doubt. Other centers of manufacturing existed, but these were no threat, and certainly no match, for Japan's industrial might.

Over the past 20 years, however, Japan has been challenged. Although it remains the undisputed leader, it now faces growing competition from the so-called Four Tigers of East and Southeast Asia: South Korea, Taiwan, Hong Kong, and Singapore. Among these, populous and productive South Korea is a formidable industrial rival. Major manufacturing districts export products ranging from automobiles and grand pianos to calculators and computers: one is centered on the capital, Seoul (18 million people), and the two others lie at the southern end of the peninsula, anchored by Pusan and Kwangju, respectively. Should the Koreas be united, the combination of the North's heavy industries and the South's major manufacturing would create a formidable industrial power just a few miles from Japan. As it is, South Korea is a growing challenge to Japan's hegemony.

To the south lies Taiwan, another growing industrial power along the Pacific Rim. The island is neither large nor populous (21 million people, compared to 46 million in South Korea), but it produces prodigiously. Taiwan's economic planners in recent years have been moving away from labor-intensive manufacturing toward high-technology industries, thus meeting Japanese competition head-on. Personal computers, telecommunications equipment, precision electronic instruments, and other high-tech products flow from

Taiwanese plants, evincing the advantages of a skilled labor force and reducing the need for massive raw material imports from far-away markets. The capital, Taipei (7 million people), is the focus of the country's industrial complex, situated on the northern and northwestern zone of the island.

Just a trading colony four decades ago, Hong Kong exploded onto the world economic scene during the 1950s with textiles and light manufactures. The success of these industries, based on plentiful and cheap labor, was followed by growing production of electrical equipment, appliances, and other household products. Site limitations constrict crowded Hong Kong, but situational advantages have contributed enormously to its fortunes. The colony became mainland China's gateway to the world, a bustling port, financial center, and coveted prize. In 1997, China will take over the government of Hong Kong from the British, and one of the world showplaces of capitalism will come under communist control, barring another democratic revolution in China. The future of this third tiger is in doubt.

The industrial growth of Singapore can be attributed in considerable measure to geography. Strategically located at the tip of the Malay Peninsula, Singapore is a small island populated by under 3 million people, most of them ethnic Chinese. There are also Malay and Indian minorities. Forty years ago, Singapore was mainly an *entrepôt*, or transshipment point, for such products as rubber, timber, and oil, but today, the bulk of its foreign revenues come from the export of manufactures and, increasingly, high-technology products. Singapore is also a center for selling services and expertise to a global market.

CHINA'S FUTURE

Could a Meiji Restoration or another kind of economic and industrial transformation take place in China, as it has among other nations of the Pacific Rim? It has already begun, but its outcome will be different. Japan had none of the domestic resources China has, but also none of the social and cultural baggage. Japan's new rulers could turn an entire, virtually homogeneous nation (Japan *is* a nation-state) into a battalion for expansion and later for economic supremacy. Japan had no land boundaries to worry about; no minorities, other than the few remaining Ainu; no cultural impediments to regimentation. In one way or another, the lives of virtually all of Japan's people (then 30 million, now 125 million) were and are affected by the military and economic campaigns of the state. In China, things are very different. China has to move more than ten times as many people into the twenty-first century, it must concern itself with tens of millions of minorities, it has contentious land borders, and, unlike insular Japan, it has neighbors that could pose problems in the future. If there is one Eurasian country that must be concerned over the possible course of events in China, it must be Russia!

Historical geographers of the future will nevertheless look back upon the 1990s as the time when China embarked on its path toward superpower status. If China does not fracture under the stresses of its regionally uneven economic growth—if its Pacific Rim provinces do not try to free themselves from Beijing's stultifying control—then only time will stand in the way. China's road will be longer, but China has the technical and intellectual ability, the productive capacity, and the domestic resources to succeed. Its dependence on foreign innovations and equipment will lessen, and the size as well as the strength of its economy (already the fourth largest in the world) will enhance its power position in a critical realm of the globe.

China is a nuclear power already. In mid-1993, the Chinese reminded the world of their disregard for the opinions of the international community by resuming underground nuclear testing. China's leaders insist that China has no military ambitions, but the modernization of the army and air force, and the expansion of the navy, have been under way since the end of Mao's era. It is not surprising that a country that counts among its neighbors a nuclear-empowered Russia, a renegade North Korea, and a missile-armed Pakistan would want to enhance its military preparedness. China, it should be noted, is a major weapons producer and exporter, and sales of arms to such countries as Iran are of concern to Western countries. In this, as in all else, China tends to act in what it perceives to be its self-interest, not in terms of the views of others.

Which brings us to the immediate concern, and back to APEC. Successive U.S. presidents during the past 20 years have sought, in various ways, to influence China's human rights practices. China was and is no dissident's paradise, before or after Tienanmen Square. Its prisons have been described as medieval. Freedoms Americans and other Westerners take for granted are not available to the great majority of China's 1.2 billion people. During the APEC meeting in Seattle late in 1993, President Clinton sought to press this issue on China's President Jiang Zemin, but with little apparent success. Earlier in the year, he had reluctantly extended China's Most-Favored-Nation (MFN) status.

The difficulties faced by the United States in its relationships with China came into sharper focus during 1994. The MFN question produced a sometimes bitter debate, in which President Clinton was reminded of his earlier pledges regarding the linkage of China's MFN status with its treatment of dissidents, while the national interest seemed to dictate a different course of action. What would be gained by increasing China's isolation and enlarging the political distance between two key powers on the Pacific Rim? Almost certainly, the fate of China's political prisoners would not improve. Prison conditions were not likely to improve either. The courageous opponents of China's harsh regime, both within and outside China, would lose visibility as well as contacts. Those who opposed continuation of China's MFN status argued that morality came first. Those who favored it pointed to the substantial improvement of the situation in

China generally, a progress that might well be reversed, and warned of job losses in the United States as well as other affected economies, notably that of Hong Kong.

In the end, Clinton not only sustained China's MFN status, he decoupled the issue from the dissident question as well, thereby forestalling another rancorous debate in ensuing years. This decision followed a disastrous visit by Secretary of State Warren Christopher to China, where the U.S. business community in a public meeting revealed the depth of its disagreement with the U.S. administration on this matter.

Through all of this, China displayed an almost complete disregard for American viewpoints, harassing and arresting dissidents at will. By a remarkable coincidence I happened to be in a bus on the highway between Tianjin and Beijing near midday on Thursday, March 31, 1994, when we were slowed and then stopped near a rest area. Several dozen uniformed police and what appeared to be plainclothesmen and their cars had blocked the two lanes in the opposite direction, and someone was being put in one of the cars. As it turned out, I apparently witnessed the arrest of the prominent dissident Wei-Jiangshing, whose return to Beijing the authorities would not allow. That the Chinese would take such action at a sensitive time served as a reminder that they pay little attention to what others may think they should do; nor, apparently, were they willing to help make President Clinton's life a little easier during the closing weeks of the debate prior to his MFN decision.

While the debate over the U.S. relationship with China focused on the human rights issue, another matter influenced it: the apparent drive by North Korea to join the nuclear club. Undoubtedly, the U.S. administration hoped that China, following the favorable action by Clinton in May 1994, would join in an effort to dissuade North Korea from pursuing its nuclear aims, even to the point of joining a United Nations sanctions policy. But China and North Korea are joined by some very strong bonds. A half million Chinese died on behalf of North Korea's campaign to reunite the peninsula by force (1950–1953). China provides North Korea with crucial resources and is its leading trade partner. And North Korea, like China, is a communist state, a rare ideological ally in the post-Soviet world. If the United States nurtured any hope that China might respond to the renewal of MFN status by abandoning its old ally, that hope was based on still another misreading of Chinese behavior.

So should the opposite decision have been taken? It is a Hobson's choice, but the lessons of historical and economic geography suggest that continuing involvement is the lesser of the evils. An open China that continues to send scholars and students to Western countries, and continues to collaborate in joint research, is a China where notions of greater freedom are likely to take root more quickly. A China with a growing middle class will demand from its leaders greater individual freedom and stronger civil rights. A China that continues to grow and interact vigorously along its Pacific Rim will eventually impose upon the dogmatists in Beijing a more liberal political system—or else China may go the way of the now-defunct Soviet empire.

If there is, from a geographic viewpoint, one regret associated with the course of events between China and the United States up to the middle of 1994, it is that the decoupling of the MFN and human rights issues was not matched by a recoupling of the Taiwan issue. That card was not played, but it might have been. Taiwan, to virtually all intents and purposes, is an independent and prosperous country, but U.S. policy toward it is constrained by China's wishes. The growing democratization of Taiwan and its economic success would seem to justify a reconsideration of U.S. relationships to this island state (or Chinese province as Beijing sees it) of 22 million people. At a time when such entities as Slovenia and Eritrea can achieve independence and Flanders and Tamil Eelam can demand it, Taiwan's case seems strong.

Too bad Mackinder could not be with us today. How would he have answered his critics now, with much of the action focused on the rimland, and the heartland in disarray? I think I can guess: "Don't count Russia out just yet," he would argue. "When their experiment with democracy ends and the Russians return to nationalist-totalitarian ways, they will have what it takes to revive their challenge for world power. What is happening in the rimland will pass; it will fracture. The heartland is forever."

The twenty-first century will prove him...what?

C H A P T E R   1 3

A UNITED STATES OF EUROPE?

For centuries, Europe has been the heart of the world. European empires spanned the globe and transformed societies far and near. European capitals were the focal points of trade networks that controlled distant resources. Millions of Europeans migrated from their homelands to the New World, as well as to newly settled parts of the Old, creating new societies from North America to Australia.

In agriculture, industry, and politics, Europe went through revolutions—and then exported those revolutions throughout the world, serving to consolidate the European advantage. Yet during the twentieth century, Europe twice plunged the world into war. In the aftermath of the Second World War (1939–1945), Europe's weakened powers lost the colonial possessions that for so long had provided wealth and influence, and the continent was divided by an ideological Iron Curtain. Resilient Western Europe's recovery and Eastern Europe's ultimate rejection of communism have been the dominant events of the past few decades.

But when we add everything together, Europe is still a world geographic realm of quite modest proportions on the peninsular margin of western Eurasia. Despite its comparatively small size, for more than 2,000 years the European realm has been a leading focus of human achievement, a hearth of innovation and invention.

Europe's human resources have been matched by its large and varied raw material base; whenever the opportunity or the need arose, the realm's physical geography proved to contain what was required. In fact, for so limited an area (slightly less than two-thirds the size of the United States), Europe's internal natural diversity is probably unmatched. From the warm shores of the Mediterranean to the frigid Scandinavian Arctic, from the flat coastlands of the North Sea to the grandeur of the Alps, and from the moist woodlands and moors of the Atlantic fringe to the semiarid prairies north of the Black Sea, Europe presents an almost infinite range of natural environments. The insular and peninsular west contrasts strongly with the more interior, continental east. A backbone laden with raw materials extends across the middle of Europe, from England eastward to Ukraine, yielding coal, iron ore, and other valuable minerals. This diversity is not confined to the physical makeup of the continent. The European realm contains peoples of many different cultural-linguistic stocks, including not only Latins, Germanics, and Slavs, but also numerous minorities such as Finns, Hungarians, and various Celtic-speaking groups.

This physical and human diversity has led to the strong and distinct development in Europe of an intricate mosaic of many countries.

And what a jumble of countries that mosaic of Europe is today! The large ones (by European standards) are familiar enough: the United Kingdom, France, Germany, Italy, Spain. But some of the smaller ministates may not be: Malta, Luxembourg, Liechtenstein. And then there are specks of land with boundaries, but with few of the trappings of real states: Andorra, San Marino, Monaco. Europe even has a leftover colony, Gibraltar, at the western entrance to the Mediterranean. Europe looks like a jigsaw puzzle.

Cultural boundaries—ethnic, linguistic, and religious—fragment Europe even more than the physical boundary mosaic. Celts in the British Isles fight a rearguard action to save what is left of their ancient culture. Catholics and Protestants in Northern Ireland fight as though the sixteenth century never ended. Christians and Muslims are destroying what remains of Yugoslavia. Europe has long been a crucible of culture, but also a cauldron of conflict.

WHERE IS EUROPE?

Silly question, it might seem. Europe, of course, is Britain and Scandinavia and Greece and Poland...but surely not Belarus or Moldova?

Welcome to a long and probably endless geographic debate. The eastern boundary of the European geographic realm has been the subject of argument for many years. At present, there are three options. One places the eastern boundary of Europe along the eastern borders of Poland, Slovakia, Hungary, and Romania so that the ex-republics of the former Soviet Union are not part of it. A second view holds that the appropriate European boundary coincides with the eastern borders of Latvia, Lithuania, Belarus, and Ukraine. On this basis, the eastern border of Europe is the western border of Russia. And the National Geographic Society promotes a third option: its Atlas has a line running down the crest of the Ural Mountains in Russia, suggesting that western Russia is part of Europe, but eastern Russia is not.

Who's right? If you ask the people at *Jeopardy*, the popular television game show that has "geography" as one of its categories, it's the second one. On the November 30, 1993, program, the contestants were asked what is the "largest country in Europe." The correct answer, which nobody got right, was Ukraine.

Actually, that does appear to be the best option. There's little justification for drawing a boundary between Poland and Lithuania or between Romania and largely Romanian Moldova. On the other hand, Russia is in a class—and realm—by itself, three times as large as all of Europe, twice as populous as Europe's largest country. For once, the National Geographic Society has it wrong.

Does it matter? Absolutely. To be a part of Europe means that a country can hope for access to Europe's many international organizations, for representation on the Council of Europe, for mutual security, and for many other advantages resulting from cooperation among neighbors. The name Europe today stands for much more than a continent or a geographic realm. It also represents international progress and opportunity. Few countries would forgo the chance to join. Eventually, despite Norway's recent decision not to join the Union, Europe's most devout neutralists, the Swiss, may be the only people left who prefer to remain outside the unifying European mainstream. "Switzerland?" said a colleague in response to a question at a postelection seminar in Paris in mid-1994. "Well, the Swiss aren't in Europe, so they feel they can do as they please, and now they're banning heavy European trucks from their Alpine roads."

What? Switzerland isn't in Europe? His remark emphasized the way Europe has been redefined. It's no longer just a geographic locale. It's a functioning complex of countries. (He was right about the trucks. In February 1994, the Swiss stopped large trucks from crossing the country on their way to and from Italy because of the damage they were doing to their vulnerable mountain roads—an action they couldn't have taken as members of Europe's Union.)

LOWERING THE BARRIERS

The great saga of Europe after the Second World War has been its gargantuan effort to overcome its historic barriers and enmities and to forge a multinational union. For more than 50 years now, Europe (chiefly Western Europe) has been working to lower the obstacles to cooperation and to facilitate the free flow of people, goods, and money. This has become the world's greatest experiment in internationalism (or, to use the geographic term, *supranationalism*). Along the way, many lessons have been learned, and these have served other countries elsewhere in the world also trying to establish supranational associations. By the mid-1990s there were more than three dozen such organizations in the world, most with subsidiaries raising the number still higher. Russia's CIS and America's NAFTA are more recent (and less mature) examples of this now-global process.

For all its success, Europe remains a strongly divided realm. There are cultural chasms between Scandinavian and Mediterranean Europe. Western and Eastern Europe still are worlds apart. Europe has rich countries and poorer ones, Euro-enthusiasts and doubters, federalists and nationalists. Twice in the twentieth century, Europe plunged the world into war. Twice the combatants came out of the conflict saying "never again." In Yugoslavia, Europeans had the chance to prove that they meant it, but they failed. In Europe, never say never.

Still, what has happened in Europe is unparalleled in the world today. It all began modestly enough: even before the end of the Second World War, leaders-in-exile of three countries—the Netherlands, Belgium, and Luxembourg—began meeting in London to plan common economic policies. Lowering their barriers to the movement of raw materials, products, and people, these leaders knew, would help during postwar reconstruction. And so the name *Benelux* came into use in England in 1944, and the initial step toward wider supranationalism was taken.

After the war ended in 1945, the Benelux Agreement began to be implemented. Other countries watched the experiment with interest, including the United States. The notion of a wider cooperative sphere in Europe was one of the tenets of the Marshall Plan, a huge American aid effort for a recovering Europe proposed by then Secretary of State George C. Marshall in a speech delivered at Harvard University on June 5, 1947. The United States was concerned that communist subversion would derail Europe's recovery, and European countries were asked to propose a giant self-help program to be funded by the United States. Sixteen European countries, later joined by West Germany, presented a plan under the so-called Organization for European Economic Cooperation (OEEC), and the U.S. Congress approved and funded it to the tune of $12 billion (about $120 billion in present dollars).

During the four years of the Marshall Plan, Western European economies grew rapidly, and the communist threat receded. Before long, the OEEC gave rise to other cooperative programs. In 1949, the Council of Europe, still only a deliberative body but regarded as a forerunner to a European parliament, was created. In 1951, a European Coal and Steel Community came into being. Bit by bit, the internal economic barriers of Europe were lowered.

But the divisiveness that has plagued Europe seemingly forever again raised its head. The Marshall Plan was phased out, and it was up to Europe's nations to continue what had been set in motion. Would the 17 countries continue their drive toward greater cooperation? The answer was a qualified yes. Qualified, because while the core powers of Western Europe wanted to continue, the British had growing reservations. The British saw their future linked more closely to the Commonwealth, and they did not want to risk those ties by joining a unified Europe. And so, when the European Economic Community (EEC) was established in 1957 by the Treaty of Rome, Britain was not a part of it. The EEC called itself the Inner Six, consisting of France, West Germany, Italy, and the three countries of Benelux.

The EEC began operation in 1958, and in the following year the British-led Outer Seven, or European Free Trade Association (EFTA), was created, a dispersed amalgam of the United Kingdom, the three Scandinavian countries, the two mountain countries (Switzerland and Austria), and Portugal. This group of states, with its limited resource base and restricted purchasing power, was no match for the EEC, but its formation underscored the divisions still afflicting Europe. As it happened, the advantages of EEC membership soon overpowered the EFTA countries, and they began to apply for EEC membership in the early 1970s. By 1973, the Six had admitted the United Kingdom, Denmark, and Ireland; the EEC had became The Nine.

As the EEC grew larger, the rewards of membership increased. Continual negotiations in many spheres produced agreements on the

EUROPEAN SUPRANATIONALISM

Original Common Market (EEC)

The EU Fifteen

The EFTA

Nonmembers

2000 Possible Date of Joining the EU

The European Community is the vanguard of supranationalism in this realm. Other countries, from the Baltic states to Turkey, wish to join.

removal of customs duties among members and on the erection of common external tariffs against imports. In EEC airports, formalities for intra-EEC travellers were simplified. In 1981, Greece joined the Community, followed in 1986 by Spain and Portugal, expanding the organization to 12 countries.

In recognition that The Twelve formed more than an economic union, EEC was shortened to EC, European Community. Before long, this name was abandoned in favor of the present one: European Union (EU), which became the official name through the Maastricht Treaty of 1991.

One reason for these changes lay in the political sphere. The Council of Europe had evolved into the European Parliament, to which participating states sent elected members for the first time in 1979. Strasbourg, located in the northeast corner of France, was the unofficial capital of Europe. (Its province, Alsace, was part of Germany and carries German imprints on its cultural landscape.) It was agreed that elections would be held every five years and that the European Parliament would expand as new members were admitted to the union. In 1979, the European Parliament had 410 members; by 1994 that number had grown to 567 and was poised to grow still further with the admission of Austria, Finland, and Sweden in 1995.

I had the good fortune of being able to witness the run-up to the June 1994 European parliamentary elections and the elections themselves, which followed shortly upon the D-Day commemorations in Normandy, where I did a segment on *Good Morning America*. What I saw and heard was something other than what the architects of a united Europe had in mind, which was to forge a European legislature where EU issues would transcend national issues within member states. During the campaigns in Britain, Ireland, France, and the Netherlands, very little was said about Union issues, and the same seemed to be true in the other EU countries. Rather, the debates centered on domestic concerns, and the voters used their ballots to punish governing parties for failing to keep their promises. The voters also expressed considerable skepticism toward the EU and a reluctance to get more deeply involved in the venture. Voter turnout was uncharacteristically low. From an EU standpoint, the only bright spot was the referendum on EU membership held in Austria. There, voters approved the idea with a two-thirds majority.

Applicants for membership seemed to be more enthusiastic toward the EU than the members already in it. EU leaders predicted that the Austrian vote would have a favorable impact on referendums to be held in Norway, Sweden, and Finland. They were wrong about Norway, but Finland and Sweden voted to join.

For member governments, the periodic European parliamentary elections are not an unmixed blessing. Major parties of member countries are awarded seats in the European Parliament on the basis of their performance in national elections. About the last thing the embattled John Major's Conservative government in Britain needed was a demonstration of its unpopularity with British voters, but that's what it got. The Tories suffered their worst defeat since coming into office, getting only 26 percent of the vote (against Labor's 45 percent) and losing almost half their seats in the European Parliament. This meant that the Conservatives, who remain in power until they must call a general election in 1997, are outnumbered, 62 to 19, in the European Parliament. Spain's Socialist Party, led by Prime Minister Felipe Gonzalez, got a similar rebuke, winning only 22 out of Spain's 64 seats.

While other national leaders fared better, including the new Italian conservatives and the German Christian Democratic Union led by Helmut Kohl, it is clear that the elections for the European Parliament pose a problem for governing parties. In Britain, I heard Conservative Party candidates tell the voters that perhaps the time had come to reconsider the Tories' position regarding the whole EU question. Polls continued to indicate that a majority of the British public is opposed to EU membership, but the government had not approved a referendum on the issue. But the June 1994 elections also revealed what seems to be an increasing disillusion with the European unification venture: results in France and Denmark, where avowedly anti-Europe movements did well, and the election of candidates who promised that they would slow the march toward union once they were seated in Strasbourg. Add to this the weak turnout, and it is clear that the next steps on the road to a unified Europe will not be easy ones.

SEATS IN THE EUROPEAN PARLIAMENT
By Member Country, 1994

Member	Population (Millions)	Seats
Belgium	10.1	25
Denmark	5.2	16
France	57.8	87
Germany	80.5	99
Greece	10.3	25
Ireland	3.6	15
Italy	58.1	87
Luxembourg	0.4	6
Netherlands	15.3	31
Portugal	10.3	25
Spain	39.1	64
United Kingdom	58.2	87
Totals	348.9	567

ECONOMIC AND POLITICAL GEOGRAPHIES

Why, with countries still clamoring to join the Union, with Union-wide rules in effect, and with Parliament in place, has the post-Maastricht period been so difficult for Europe? In large measure the answer lies in economics. Europe made great headway while European economies were booming. In the aftermath of the Marshall Plan, and in a world still largely dominated by European influence, national economic and political interests seemed to converge—especially while the Iron Curtain still shrouded the intentions of a powerful adversary. Millions of immigrants entered Europe from Turkey (most of whom went to West Germany), Algeria (to France), and other non-European countries to take jobs often unwanted by Europeans themselves. As long as economies grew and unemployment was not a problem, these immigrants were welcome, despite occasional friction in the growing ethnic neighborhoods of larger European cities.

As recently as the 1980s, Europe seemed to be on a roll not soon to end. European products secured worldwide markets. Incomes in several European countries outpaced those of Americans. European modernization of airports, highways, railroads, and other facilities outstripped the rest of the world. It seemed that Europe, unified this time, would become a power core that might once again dominate the world.

When the Iron Curtain fell and when the Berlin Wall was demolished, reuniting West and East Germany, Europe's wildest dreams seemed on the verge of reality. All 12 member states signed the so-called Single European Act, the next step on the road toward unification. That act specified as many as 279 discrete goals to be achieved throughout the European Union. From agricultural policies and practices to labor rules and rights, the EU would set consistent standards.

When Europe's Twelve met in the Dutch town of Maastricht toward the end of 1991, they were there to create the instrument that would enable the union to take the next step. The goals of the Single European Act would be implemented by December 31, 1992. There would be a common currency before the year 2000. Through a complicated system of measures, the richer countries of the European core would help the poorer ones on the periphery (Portugal and Ireland would be major beneficiaries). And the last barriers to movement within The Twelve would disappear. For example, a person born and living in one member country would be able to live and work in any other EU country without bureaucratic interference. It would be as easy as moving from Michigan to Florida.

A truly united Europe now seemed to be in sight, and the newly coined European Union was the name to rally the member nations. But even as newspaper headlines trumpeted Europhoria, a word coined to describe the heady atmosphere, trouble loomed. The trouble was not political: it was economic. Quite quickly, EU countries' economies went into recession, the worst downturn since the war. Germany's reunification involved economic and social problems far worse than had been anticipated. Social unrest plagued many cities; protests against provisions of the Single European Act by French farmers produced violence and disruption. Immigrants found themselves the targets of attacks, and murders of Turks in Germany shocked the world. Extremist movements arose, and political parties with antiforeigner platforms attracted substantial support.

As Europe's economies faltered, unemployment rose and became a major issue. In Spain in mid-1994, unemployment stood at 23 percent; in Ireland it was 18 percent. Everywhere else it stood above 10 percent except in Germany, Portugal, and Luxembourg—Portugal because of its EU benefits, which helped put thousands to work on infrastructure improvement. These were the conditions under which the 1994 European parliamentary elections were held, and it is not difficult to see why enthusiasm for the concept of a united Europe was on the wane. Indeed, newspaper headlines now took to worrying about emerging Europhobia.

Europe thus confirms an old rule: political progress in a supranational context comes more easily when economic times are good; during bad economic times, success is defined by hanging on to the gains made. If Europe can traverse its current troubles without losing ground, there may still be a federation in its future.

POLITICS AND SECURITY

When Europe emerged from the Second World War, communism posed an immediate threat to free institutions and open economies. The Marshall Plan was designed to revive economies and promote regional integration, but there was another crucial need: security against what appeared to be an aggressive Soviet Union. The Iron Curtain, the Berlin Wall, and other manifestations of Soviet hostility required a counterweight. The North Atlantic Treaty Organization

NATO MEMBERSHIP

The current membership of NATO is as follows:

Belgium	Germany	Luxembourg	Spain
Canada	Greece	Netherlands	Turkey
Denmark	Iceland	Norway	United Kingdom
France	Italy	Portugal	United States

European Command Headquarters: Casteau, Belgium
Atlantic Ocean Command Headquarters: Norfolk, Virginia
Channel Command Headquarters: Northwood, United Kingdom

(NATO) linked the United States and Canada to 13 European states in a military alliance, now consisting of 16 countries, that promised collective defense if a single member were attacked.

Military alliances do well when the threat they address is clear and present. When that threat recedes, the parties tend to start squabbling. In the case of NATO, the first 20 years were purposeful and productive. But when the Soviet threat seemed to diminish during the late 1980s, NATO faced some difficulties. Unlike the EEC, NATO included the United States, and in some European countries, notably France, political discourse revealed a dislike for a perceived American domination of the organization. Old rivalries among the British, French, and Germans reemerged. NATO membership also caused political problems for governing parties in Greece and Turkey. Widespread opposition in Europe to the Indochina War had the further effect of devaluing NATO membership, in the view of many Europeans.

NATO successfully adapted to these changing circumstances, but then came its greatest challenge: the collapse of the Soviet Union. Did the end of the communist threat mean that NATO had outlived its usefulness? Was this the time for an all-EU military organization, without the United States or Canada?

This latter notion gained substance not only through the long-existing Conference on Security and Cooperation in Europe (CSCE), but also, and more significantly, through the Western European Union (WEU), the defense arm of the European Union. Enshrined by the Treaty of Maastricht, the WEU has already acted on its own terms, irrespective of NATO's wishes.

One of these terms involves expanding the membership of Western European security organizations in Europe and, conceivably, even into the ex-Soviet states they were formed to contain, including Russia. Not long after the collapse of the communist empire, several former communist countries expressed a desire to join the very NATO whose future some in the West saw as being in doubt. Should Hungary, Poland, Ukraine, and other countries of the former communist sphere reconstitute NATO and give the organization a new purpose? From Russia came a resounding nyet: Moscow made it clear that full membership in NATO for countries on or near its borders (especially in its near-abroad) would not be acceptable. The EU's Western European Union, however, did not wait. Early in 1994 it admitted as "associate partners" nine East European countries, including the three Baltic states. The Russians were not pleased about this and proclaimed their right to intervene in former Soviet republics to defend the interests of Russians still living there.

Moscow also made clear its desire for a veto over NATO expansion into Eastern Europe. In any association with NATO, the Russians wanted to be treated as the preeminent power among former NATO adversaries and the only one with a still-functional nuclear arsenal. All this created novel problems for NATO, and in late 1993 President Clinton traveled to Brussels to declare that the United States opposed expanding NATO into Eastern Europe. Among the reasons he cited was the possibility that a political setback for democracy in one or more of the Eastern European applicants could bring a totalitarian regime into NATO's fold.

In place of full membership, NATO proposed a plan referred to as the Partnership for Peace. As many as 20 countries had expressed an interest in an association with NATO; the Partnership for Peace plan would establish "good neighbor" accords with all those governments that were interested. Nine Eastern European states were invited to begin negotiations. Russia also was invited. And after some hesitation, Russia announced its intention to sign a Partnership for Peace accord with NATO, even without the veto rights it had earlier sought. The U.S. Secretary of State described this breakthrough as "the most important strategic innovation in Europe since the creation of NATO."

While conferences were held to determine NATO's future dimensions and functions, the organization lost a gigantic opportunity to prove its worth in the new Europe. In Yugoslavia, the transition from communist dictatorship to what many hoped would be a multicultural federal state broke down, and after a brief skirmish, Slovenia was the first republic to break away. An entire world could see what would happen next, and for a brief period there was an opportunity for Pan-European intervention through NATO's forces. Instead, NATO members bickered over policy, Germany prematurely awarded recognition to formerly Nazi-supporting Croatia, and the former Yugoslavia descended into civil war and human disaster. Europe's collective failure in Yugoslavia was a serious blow to the advancing mood in favor of unification. Just as Americans during the Vietnam War saw night after night of horrible, wrenching pictures from the battlefields and villages there, so Europeans for over two years have been exposed to savagery not seen since the Second World War. And this exposure has rekindled notions of division, of incompatibility, and of nationalism at a crucial time in Europe's unification campaign. How secure, many Europeans wondered, would they be in a supranational Europe? Who would be called upon to douse fires of unrest should they strike Germany? Greece? Italy? The road to Maastricht had been exhilarating, for all its bumps and deviations. The path beyond would be a far greater challenge.

TROUBLED FUTURES

There's no question about it: the decade of the 1990s, after a heady start, has been a troubled one for The Twelve, and for Europe generally. European economies were declining and deficit-ridden, unemployment was high, public dissatisfaction was rife. Pessimism pervaded virtually all reports on Europe's future, from the popular press to the scholarly literature.

In truth, Europe's much-publicized problems conceal a host of lasting accomplishments. But during the 1990s, the European Union came face to face with some of the most difficult issues arising from sustained supranationalism: monetary union and increasing political coordination. These issues touched on the most sensitive nerves: imagine a France without its franc, a Germany without its mark, a Britain without its pound! Soon after Maastricht, an agreement to keep the values of existing currencies within a relatively narrow range, as a preliminary move to a joint banknote, failed under the economic pressures in individual countries.

That failure made much news, as it came during a time when The Twelve were also considering the expansion of the EU to 16 members by 1995. If The Twelve could not agree to carry out a monetary agreement of this kind, wouldn't expansion to 16 members make success even less likely? And so the EU became embroiled in a debate over breadth versus depth. In May 1994, just before the June elections, the European Parliament voted enthusiastically for the admission of the four applicants, should their voters approve. But this served to heighten the fears of those who want to move Europe toward federal integration: widen the Union, they argued, and it will remain a shallow association. The British, on the other hand, always fearful of just such federalism, saw enlargement as insurance against this. And the British have their allies in other EU countries. The yielding of authority by member governments to the European Parliament in Strasbourg is a slow, halting process. The European Parliament may, for a long time, remain a rubber-stamp assembly for actions by member states, not a true decision-making body for a supranational Europe.

While the political wrangling continues, an ever growing number of EU organizations continues to coordinate life in the member states, and these organizations, to be sure, are in no danger of failure. In a way, the situation resembles that of the United Nations. There, the General Assembly's quarrels and the Security Council's disputes make the news, but behind the scenes the U.N.'s agencies—the Food and Agriculture Organization (FAO), the World Health Organization (WHO), the High Commission for Refugees (UNHCR), and many others—provide crucial international services and relief. In Europe, the European Coal and Steel Community, the European Atomic Energy Community, the Charter of Paris, and many other multinational organizations, some confined within the EU and others extending to nonmembers, are the real achievement of 50 years of supranationalism. Arrive at European airports, cross European Union boundaries, do business in the EU, and the results and rewards of supranationalism will immediately present themselves.

So is the malaise of the 1990s just another of Europe's occasional downturns? After all, Europe has gone through many ups and downs. Is the present downturn just a temporary setback?

From a geographic perspective, the answer is not encouraging. Even as Europe's internal strength and unity have improved during its half century of progress toward unification, Europe's regional geography has markedly changed and Europe's situation in the world at large has also been altered. Here are some factors to consider.

THE CHANGING MAP

When the original EEC was created, it was expected to grow, and the hope was that some day it would encompass all countries west of the Iron Curtain. The potential for supranationalism in Europe ended there; Greece might someday join, but no Eastern European country.

Now, the map of Europe includes not only the old Eastern

Europe of Poland, Hungary, and Bulgaria, but also the new Eastern Europe of Latvia, Belarus and Ukraine. Already, countries of the old Eastern Europe have applied for membership in the EU, because they recognize the advantages of involvement.

But many of the stipulations of the Single European Act, and the terms of the Treaty of Maastricht, cannot be met by these aspiring members. It was difficult enough for The Nine to become The Twelve, as it was agreed that richer members would support poorer members through transfers of funds. Should Poland, the Czech Republic, Slovakia, Hungary, and Romania join, that practice would no longer continue. In 1994, as noted earlier, voters in Austria, Finland, Sweden, and Norway were asked to decide on their countries' membership in the EU. In Austria, where polls had indicated a close vote, the margin in favor of membership was unexpectedly large. Finland and Sweden, too, reported comfortable majorities in support of participation. But Norway's voters declined, despite a government-sponsored campaign urging approval. Polls showed that the majority of Norwegians worried that joining the EU would dilute Norway's control over its two vital resources, the North Sea oil and gas fields and the rich fishing grounds within its EEZ. On January 1, 1995, therefore, Europe's Twelve became The Fifteen, and now the arena of expansion would shift eastward.

And there, the challenges will be even greater. Applications from Malta and Turkey are pending, and Poland, the Czech Republic, Slovakia, Hungary, Romania, and Bulgaria dream of membership by 2000. After painstakingly slow and deliberate admissions between 1957 and 1991, an avalanche of new membership demands now faces the EU—and with it new risks.

One of these risks involves Turkey. The Turks, loyal NATO members from the beginning, want EU membership and have suggested the year 2000 as a date of admission. But the price of EU growth is the potential for greater conflict. The old (indeed, ancient) hatchet between Turkey and Greece has never been buried. When

Greeks were polled regarding the admission of Turkey, more than 98 percent of respondents opposed admission.

In Western Europe, however, Turkey's admission is seen in another light. There, the hope is that Turkey's membership in a future EU would open doors to the secular Muslim world, especially in Turkestan, where Turkish influence is strengthening. All that, however, lies in the distant future. Confronting Turkey's application for membership in the EU will provoke a debate the likes of which the EU has not yet seen.

CHANGING RELATIVE LOCATION

Europe's relative location has also changed. When European powers embarked on their campaigns of colonization and exploitation, Europe's situation in the world was in itself a resource: North Atlantic trade routes beckoned, North African colonies lay open to France, Spain, and Italy; the Suez Canal afforded access to the Indian and Pacific Oceans and their profitable shores. But look at Europe now: flanking it to the south is an Islamic world in turmoil. Algeria as a French colony had more than 1 million French residents, and after more than a century of transculturation, France left behind a secular Muslim state anchored in the Francophone community. While oil and gas prices were high, Algeria's economy prospered, and France was Algeria's chief trading partner. But when those prices dropped and Algeria's economy faltered, the former colony proved fertile ground for those who blamed it all on Western decadence: the Islamic fundamentalists. Local elections late in 1991 proved the popularity of the Islamists, who seemed poised to win the January 1992 elections. The prospect of fundamentalist Islamic regime in Algiers caused the government to cancel the 1992 vote, and this plunged Algeria into crisis. President Mohammed Boudiaf was assassinated in June 1992 after the government had banned all nonreligious activity at Algeria's 10,000 mosques.

By early 1995, Algeria was locked in a civil war reminiscent of

Iran in the late 1970s. More than 30,000 people had been killed by Islamic extremists and pro-government death squads; foreigners were specifically targeted and embassy personnel, oil-facility technicians, professors, journalists, and priests were among the victims. The first direct involvement of France came in December 1994, when Islamic extremists hijacked an airliner they planned to blow up as it overflew Paris. Their plot was foiled at Marseille, but it evinced the fundamentalists' plan to bring Algeria's crisis to French soil.

The prospect of a bitterly anti-Western, fundamentalist regime in Algeria is a major portent for Europe as well as North Africa. Compare Algeria's location to that of Iran: the Shah's ouster brought a radical regime to a comparatively remote country adjoined by the likes of Afghanistan, Turkmenistan, and Iraq. Algeria, on the other hand, is bordered by the long-stable, conservative Kingdom of Morocco to the west and progressive Tunisia to the east, both potentially vulnerable to fundamentalist campaigns. For Europe, a militant and extremist Islamic core across the Mediterranean would pose critical problems, not the least of which would be a massive migration of political refugees. Already, observers call the Mediterranean the "Rio Grande of Europe"; but what lies in store may dwarf the flow of migrants across our southern border.

Thus Europe's relative location is changing. To the east lies a Russia whose economy, to put it in a positive light, is in a transition whose outcome is uncertain. Russia's future as a stable and coherent state is by no means guaranteed, and a breakdown of order in the Russian Federation would have an incalculable impact on Europe.

And to the west, the Atlantic Ocean no longer is the world's leading trade route. And Europe lies about as far from the dramatic economic developments on the Pacific Rim as one can get. Geography, so long a European ally, is now moving its favors elsewhere.

SOCIAL GEOGRAPHIES

Another growing issue to be confronted when the EU expands has to do with human rights, specifically, the treatment of minorities. The Maastricht Treaty's rules regarding human rights are quite specific, and even non-Greeks in the EU argue that Turkey has not made sufficient progress in this area to qualify for membership.

The human rights issue, however, is not just a matter for the future. While Turkey itself may not have an adequate human rights record, Turkish workers and their families living in Germany have been subjected to a rising tide of violence during the 1990s. German mobs have attacked Turkish immigrant hostels, and Turks have been killed, including children. At first, the attacks seemed to be isolated incidents perpetrated by individual extremists, but soon a pattern of organized assaults emerged. Neo-Nazis, skinheads, and other radicals tore into Turkish communities, often without interference from watching, sometimes applauding crowds. Immigrants and visitors from other parts of the world, especially Africa, also were threatened and attacked.

Those who saw this as merely a fringe movement were contradicted by German election returns. Antiforeigner, extreme right-wing sentiment manifested itself in local elections. Nor was Germany the only country to experience such violence and right-wing political reaction. In France, anti-Algerian (and other North African) sentiment led to violence, and again a right-wing national political party showed the strength of this radicalism.

Since the great majority of Turkish, Algerian, and other immigrants to EU countries live in the cities and larger towns, where they have established ethnic enclaves, this violence had its greatest impact there. During a 1993 trip through 13 European cities, I was dismayed at the change from just four years earlier, when I had visited seven of them. Now I saw swastikas painted or posted on walls in every one of those cities, even where wartime Nazism had been felt most severely, as in Amsterdam and Paris. Marches and counter-

THE IMMIGRANT INVASION

Europe in modern times has sent millions of its inhabitants to populate the Americas, Australia, and other overseas realms. Where they overwhelmed local communities, the white settlers from Europe created new societies in the European mold. Even where they remained in the minority, as in South Africa and Algeria, they drastically changed their new homelands.

Now Europe is experiencing an immigrant invasion itself. In recent decades, immigrants have come to Europe—especially to Western Europe—in large numbers. By the mid-1990s, about 15 million immigrants were living in Western Europe, nearly 9 percent of the region's population. The great majority came from the former colonies with many arriving from comparatively nearby countries, such as Turkey and Algeria, to take jobs that became available during Europe's industrial boom of the 1960s and 1970s. Others came from more distant one-time colonies, such as India, Indonesia, Angola, and Suriname, exercising their right to do so as subjects of the colonial power. And since the fall of communism in 1989, hundreds of thousands of migrants have entered the region from Eastern Europe. Almost all, whatever their origin, settled in Europe's great urban areas, where jobs and other opportunities were located.

When Europe boomed, these immigrants found employment and were generally welcomed. But when Western Europe's economies slowed after 1980, the newcomers were less welcome. Opposition to uncontrolled immigration arose. Competition between the immigrants and Europeans for scarcer jobs increased. Social problems in the cities intensified. For the immigrants, Europe did not prove to be the melting pot that Europeans had found in America. Charges of discrimination were made against European governments, and many European cities with ethnic communities, where the immigrants mostly cluster, found themselves with problems for which they were not prepared.

Major European cities, including capitals such as Paris and Amsterdam, contain large, cohesive neighborhoods where the immigrants have implanted some of their own culture. France's leading cities (not only Paris, but also Lyon and Marseille) have many suburbs where street signs are in Arabic, where Islam rules, and where the atmosphere is that of urban Morocco, Tunisia, or Algeria. The Turkish imprint on many German cities is similarly strong, although of the 2-million-plus Turks, only a few thousand have been permitted to become German citizens. In Amsterdam, tens of thousands of Surinamese immigrants have changed the face of the city, whose social geography has altered rapidly as large numbers of white residents have moved to outer suburbs. Such changes cannot occur without difficulty, and Amsterdam suffers from increased crime and related problems.

Thus Western Europe, itself a patchwork of societies and traditions, faces the need for new adjustments and the reality of a new social map. The transition is not going smoothly in many quarters, however, as reactionary political movements gain strength, agitation to sharply limit immigration intensifies, particularly in Germany, and acts of violence against newcomers multiply. Europe has been described as the most international of realms; almost overnight, it has also become the most intercultural.

marches, demonstrations and protests filled the streets on many a summer night. But the ultimate cause was economic: the evidence of joblessness was everywhere. On the day I reached Antwerp (where hate graffiti seemed especially prevalent) the newspapers reported an unemployment figure of nearly 15 percent for Belgium as a whole.

The social geography of EU countries is changing in another significant way. The population is aging rapidly, putting pressure on governments to fund growing pension schemes. But Europe in the mid-1990s is going through its worst recession since the 1930s, and money cannot now be set aside for these purposes. Many European societies are near zero growth demographically, a goal to which some population policy makers aspire. One consequence of zero population growth (ZPG), however, is a top-heavy population pyramid whose aged must be supported, in their retirement, by the productivity of the young. For the fewer young workers, this means greater burdens, including higher taxes. But higher taxes inhibit economic growth. Once again, Europe faces a problem it has not confronted in this form previously, and it is far from clear just how it will be overcome.

And so there are several serious liabilities to weigh against Europe's assets and achievements. Europe's greatest accomplishment of the 1990s may well be to consolidate, and not to retreat from, the level of supranationalism it reached when the Single European Act was signed. Further expansion of the EU, without watering it down, may have to await an economic upturn, however remote that prospect currently is.

THE DISUNITED STATES

Whatever the outcome of Europe's current struggle to redefine, reconstruct, and rename itself (it has gone from OECD to EEC to EC to EU, so another name may well be in the offing), it is clear that there will be no United States of Europe in the foreseeable future. When the need for mutual trust and cooperation was greatest, in economically difficult times, Europeans once again proved their fractiousness.

This is not to say that the remaining obstacles to true union are minor. Even before the recession of the 1990s struck, the three giants of Western Europe revealed the depth of their continuing mistrust. That was before Germany was saddled with its reunification: the West German economy boomed, and the French and the British worried about German domination of a future European Union. Next, the British decided that the European Monetary Union was against their interests, at least insofar as the Exchange Rate Mechanism was concerned—and they opted out of it. Most recently, the French forced a modification of common agricultural policy after French farmers took to the streets and highways to demand protection. All this had the effect of eroding the gains made toward a more effective union, and it rekindled nationalisms European leaders want to moderate.

The two biggest threats to a more unified and stable Europe, however, may lie in the devolutionary forces now affecting so many of its states and in the potential for wider conflict arising from the strife in the former Yugoslavia. These devolutionary forces affect, with greater or lesser intensity, EU members Spain, France, the United Kingdom, Belgium, and Italy. Greece lies in the path of any expansion of the Yugoslavian conflict, and Italy may not be safe from it either. With the liberation of the former Soviet republics, the new Eastern Europe creates a new European frontier. At the same time it makes the old Eastern Europe a more central region, part of the heart rather than the periphery of Europe. As such, the disaster that has befallen it resonates far to the west, where latent cultural conflicts also exist.

Europe's spectacular progress toward unification took place in the shadow of the Iron Curtain, was stimulated by the shared threat of communism, and had the security of protection under

NATO. Now Europe is on its own; the remaining superpower is
on the far side of the Atlantic, the world's economic boom is on the
far side of Eurasia. In the new world order now in the making,
European supranationalism will face its toughest test yet.

CHAPTER 14

SHAPING THE NEW WORLD ORDER

So many changes affect the geopolitical world today that it is difficult to keep track of them—and even more difficult to discern long-term trends that may give us a hint of what lies ahead. Geographers keep track of new states and their borders, new subdivisions within states that sometimes reflect devolutionary pressures, and signals such as election returns and their geographic implications. But much of what is stirring the world today cannot be put on a map, at least not on a global scale. Europe's drive toward union is unfinished and now stalled. The Pacific Rim is in rapid transformation. Russian nationalism is reawakening. China's communist empire is on the move. New maps will undoubtedly be drawn someday, but what will they look like?

During the brief post–Cold War period, I have been able to visit more than 50 countries, some of them more than once, on all six inhabited continents. I found myself in the middle of an anti-Russian demonstration in Riga, Latvia, saw drugs and arms sold on the Chinese island of Hainan, felt the tension in changing South Africa, heard southern Brazilians call for freedom, interviewed Scottish leaders about autonomy. Three threads run through my notes:

1. *Demands for Democracy*

Wherever one goes and listens, the call for democracy can be heard—loudly in some democracies where the system does not deliver what is expected of it, and quietly where repressive regimes still rule. While the notion of democracy is widely embraced, definitions of democracy vary greatly. In Kenya last year I got embroiled in a vigorous debate over what Africans like to call one-party democracy. The Kenyan politician's argument was that it is quite possible to encompass a wide range of political positions within a single party. "In parliament," he said, "representatives are therefore not constrained by party affiliation, but only by what is best for their constituents." In his view, the U.S. system is in effect a one-party system, divided largely for game-playing and only slightly by ideology, designed to maintain a monopoly. "See the difference?" he wrote me after the October 1993 Canadian election. "*Their* Conservative Party was just about voted out, and went from more than 150 seats to two. *Their* Ross Perot (Preston Manning, the Reform Party leader) got votes and seats. *Your* so-called United We Stand got nearly 20 percent of the vote, but I don't see any UWS members in your Congress. What kind of democracy is that?" So, he added, don't lecture us about multiparty democracy.

Discuss democracy with Chinese politicians, and you will hear, as I have repeatedly since my first visit in 1981, that Chinese Communist Party officials genuinely believe that theirs is a democratic system. (I got the same lecture in Russia during my initial visit there in 1964.) Even after the prodemocracy movement of the 1980s, and the disastrous events of Tienanmen Square and less heralded events in other cities, this is still the party line.

Chinese prodemocracy demonstrators carted a replica of the Statue of Liberty into Tienanmen Square, but other visions of democracy do not mirror the American model. Singapore constrains democratic practice, in part by limiting individual freedoms; democracy in Mexico (or rather, the lack of it) became a major issue during the debate in the United States over NAFTA. But even in countries where little progress toward representative government or multiparty democracy has been made (Indonesia, Saudi Arabia, Zaire) the idea of democracy—however vague its definition—stirs the hopes of millions.

2. *Visions of Independence*

Perhaps independence is too strong a word; many who use it to lift their spirits would be happy to substitute autonomy, or self-determination, or home rule. But whatever form it takes, the vision of independence seems to be everywhere. Long dormant among peoples enmeshed in the networks of larger states, newly revived visions of independence are stirring nations, tribes, and clans.

As my experience in Scotland proved, the people who desire greater self-determination are not always in accord on just what they want or on ways to achieve it. I met Scottish nationalists who want nothing less than outright sovereignty, and others who would be satisfied with a greater degree of autonomy within the United Kingdom. An empty building sits ready in Edinburgh to accommodate a future Scottish Parliament, and this, it seems, a majority of Scots would like to see occupied. In Hawai'i, native Hawai'ians are not at all in agreement on what their claim to independence should embody. In Flanders, newspapers print editorials, op-ed pieces, and letters to the editor that reveal how divergent opinions on a future, separate Flanders really are.

The growing list of secessionist states emboldens even the smallest community, and once kindled, the flames of independence are hard to extinguish. When the Slovenes and the Eritreans and the Estonians can succeed, and when the Tatars and the Tamils and the Cypriot Turks can sustain a challenge politically or militarily, why not the Chechens and the Corsicans and the Abkhazians? Many of these aspiring groups may not be familiar to us now, but some will undoubtedly be members of the United Nations before long. Independence is a fever that has become a global pandemic.

3. *Revival of Religions*

The third theme that recurs wherever one goes is religious: the revival of belief systems. I use this latter term because what is happening is more than a return to the faith: it is a return to cultural basics, among which religion is one.

Almost everywhere I have gone over the past several years, the signs have been powerful, sometimes overwhelming. None of it should be cause for surprise: religions are historically reinvigorated in cycles. But it is the worldwide nature of the present revival that makes it remarkable.

In communist China, Mao Zedong's regime made a gigantic effort to eradicate Confucian thought and practice. Even today, the more liberal communist rulers have not officially reversed this objective. In my book *Geography: Regions and Concepts*, published in 1971, I wrote the following:

> Confucianism was attacked on all fronts, the Classics were abandoned, ideological indoctrination pervaded the new education, even the (nuclear) family was assaulted during the early days of communization. But it is difficult to eradicate two millennia of cultural conditioning in three decades. The dying spirit of Confucius will haunt physical and mental landscapes in China for years to come.

I should have written not that it would be difficult, but impossible. Today, Confucianism is experiencing a revival, even as more formal religions, including Christianity and Islam, are. In 1981, I saw China's largest mosque, in the western city of Xian; it was a small, walled enclosure in a cluttered quarter of town. In 1992, the mosque was a bustling, thriving establishment, spacious and well-attended. Even in China, the religious revival is under way.

What is happening in China is only one manifestation. From America's Bible Belt to India's Hindu heartland, and from Turkestan to Russia, the role of religion is resurgent—whether by rebirth, as in Islamic Turkestan and Christian Russia, or by reintensification, as in Muslim Algeria and Hindu India. We in the Western world sometimes equate religious fervor with Islamic extremism, but the religious revival of today is a broader, global phenomenon, its manifestations ranging from the doorways of abortion clinics in America to the voting booths of India.

GEOGRAPHIC FACTORS

What lies behind these apparently long-term, world-shaping trends? I have touched upon several possible factors previously in this book, but looking at the evolving new world order geographically, the following factors emerge:

1. *Environmental Forces*

The effect of increasing environmental change upon human society is as yet difficult to interpret, but there is no lack of evidence for the role the environment has played in generating political change. The failure of the Soviets' infamous Virgin Lands Scheme in Turkestan, wherein millions of acres of steppe were irrigated and fertilized to create farmlands, produced an ecological disaster. The rising tide of anger over the medical consequences (pesticides polluted the sources of groundwater, killing and disabling thousands) contributed significantly to the region's burgeoning resentment over Russian rule. Ethiopia's Marxist regime was defeated by a sequence of droughts, related to the Sahel crisis, that drove the population off the land and galvanized regional opposition. The increasing frequency and intensity of El Niño events, combined with overfishing, destroyed one of Peru's major sources of income and contributed to the malaise that helped foster the Sendero Luminoso (Shining Path) Marxist rebellion. Closer to home, the Sunbelt migration is in part a response to changing natural environments, with considerable political implications.

**EUROPE: FOCI OF
DEVOLUTIONARY PRESSURES, 1995**

Devolutionary stresses affect long-stable states as well as more recently decolonized countries. During the 1990s, Yugoslavia was dismembered by tribal warfare.

2. *Failures of the State System*

One consequence of the long-term Cold War was a so-called bipolar power balance that had the effect of concealing a major weakness in the global political system: the national state.

The dissolution of the Soviet empire and the release of Eastern

Europe from the communist yoke have brought these weaknesses to the fore, but there was earlier evidence. The European state model proved to be difficult to adapt to the conditions prevailing in the colonies where it was installed; and in Europe itself, signs of internal stress were evident. Europe's own unification movement, beginning even before the outbreak of the Cold War, is evidence that states were impeding regional growth if not bound by multinational agreements. Noteworthy as the European Union's successes may be, they are in and of themselves evidence that the state as we have known it for more than two centuries is losing its effectiveness and relevance as the key element in the global politico-geographic system.

3. *An Antiquated Boundary Framework*

The world is entering the twenty-first century with a boundary framework rooted in the nineteenth, which is a recipe for disorder. Furthermore, population growth and migratory shifts have rendered large segments of the world's boundary framework obsolete, or worse, divisive.

So long as boundary realignments remain rarities, resulting only from disastrous dislocations such as that now engulfing the former Yugoslavia, boundaries (and not only international boundaries, but internal borders as well) will continue to threaten the future stability of the world. The alternative is to reduce the barrier effect of boundaries by international agreement, but this, as the European experiment has proven, is difficult to accomplish.

4. *Supranationalist Schemes*

The idea that states should combine in regional, cultural, economic, or other alliances is not new. The ancient Greeks tried it, and so did the cities of the Hanseatic League. Now we live in an age of numerous and sometimes overlapping multinational combinations, an age of acronyms: NAFTA, EC, APEC, ASEAN, OAS, NATO.

Supranationalism, perversely, also stimulates notions of autonomy among those it affects. Scottish nationalism, I discovered, was spurred by fears that in a future united Europe, Scotland would no longer be a second-tier division in a major state, but would become a third-tier unit in a much larger framework. So the Scots revved up their democracy campaign, hoping, in the words of one of my hosts, "to sit at the table in Brussels or Strasbourg with the same rights as the Danes or the Norwegians." You can hear similar sentiments in many other parts of Europe: supranationalism is provoking separatism even as its grand design unifies a community of nations.

5. *Migrations*

Not only has the world's human population grown like a bacterial organism in an ideal environment, it is more mobile, at least in terms of absolute numbers, than ever before. Human migration streams are transforming nation-states and regions, and they are transforming political landscapes from Germany to Indonesia. Millions of people move in desperation, as refugees, but additional millions migrate in search of better economic opportunities. In Italy, the North Africanization of the Mezzogiorno south of Rome is tangible evidence of the impact such movements have on individual countries. Already, southern Italy was the poorest and least developed part of the nation. The inflow of Muslim North Africans, many of whom intended to cross Italy into the heartland of Western Europe, but who instead settled in the Mezzogiorno, has become a significant centrifugal force in Italian political geography. The reactionary Lombardy League of the North manifests the intensifying north-south split in this country. Algerians in France, Turks in Germany, Africans and Asians in Britain, Haitians in Florida—all symbolize the shrinking and internationalization of the world, and all have generated often negative political responses. In Germany and France, political polarization has a spatial dimension directly related to the enclaves of the non-European immigrants. From Kenya to Pakistan and from the Southwestern United States to

Southeast Asia, migration streams are contributing to the stresses upon nation-states.

6. *The Flow of Ideas*

Visit a remote African village, and transistor radios will be tuned to the voices and music of distant lands. Walk the streets of Saigon, and CNN flickers on television screens, received via often primitive dishes on dilapidated rooftops. No longer is it possible for even a dictatorial regime to insulate a population against external news and ideas—not even North Korea or Burma (Myanmar). Human geographers who study the diffusion of information from Western Europe into the East during a period of political transition report that radio, television, and the press contribute crucially to the resurgence of anticommunist nationalisms there. My own recent experience in Vietnam was relevant in this context. Virtually everyone with whom I was able to talk in Saigon knew of the failure of communist systems. Many expressed frustration at the persistence of communist rule in (and from) Hanoi and believed that the dogmatic northerners, as much as the U.S. embargo, were delaying Vietnam's participation in the economic boom along the Pacific Rim. Notions of democracy can now be carried to the most isolated places in interior Zaire, insular Indonesia, and remote Burma (Myanmar), and such notions often translate into ethnic fervor.

7. *The Flow of Weapons*

The splintering of Yugoslavia, the collapse of Liberia, the near-disintegration of order in Sri Lanka, and the chaotic power struggle in Somalia all remind us of the ready availability of weapons virtually all over the world. We should not be surprised: the United States has long ranked at or near the top of the list of arms manufacturers and distributors. The Somali experience is a case in point. When Somalia was a Soviet client (neighboring Ethiopia was aligned with the United States), the Soviets flooded their ally with weapons, many of them small arms to be used in challenging Ethiopian troops over the Ogaden region. Then, when Ethiopia turned Marxist and Somalia became a U.S. ally, American weapons flowed in. As a result, when the government in Mogadishu faced a political challenge, that challenge was backed up by a seemingly inexhaustible supply of firepower. After its overthrow, Somalia disintegrated into a mosaic of warring factions, and only the northwest remained reasonably stable. Now, neither the United States nor the United Nations seems able to disarm Somalia's clans, and U.S. Marines have been killed by Somalis using American arms. This is by no means a unique story: whether in Angola, or in Liberia, or in Afghanistan, or in Bosnia, the supply of weapons is inexhaustible. And their manufacture and distribution—from the factories of the United States, China, Russia, Brazil, and countless more—continues.

8. *Irredentism*

It is natural for a people to be concerned over the fate of their kin living separate from the motherland and in difficult circumstances. But when such concern becomes national policy and then strategy, it can adversely affect the stability of nations. The word for such policy is *irredentism*, and it is a potent force in the splintering of states. Hitler used this ploy to extort the Sudetenland from Czechslovakia just before the start of the Second World War. In Eastern Europe today, a potentially dangerous situation of this kind involves Hungary and the Hungarian minorities beyond its borders. In the heady days after the defeat of communism, a Hungarian leader proclaimed himself leader of all Hungarians, including those in Romania, Slovakia, and Serbia. Hungarian newspapers and other media can, in short order, arouse local anger over anti-Hungarian practices in neighboring countries. As Hungary, among other Eastern European countries, purchases weapons on world markets, concern rises over the future of the heartland of Eastern Europe. If Hungarians embark on a campaign similar to that waged by the Serbs in the former

THE NEW EASTERN EUROPE

POPULATION

- Under 50,000
- 50,000–250,000
- 250,000–1,000,000
- 1,000,000–5,000,000
- Over 5,000,000

National capitals are underlined

0   100   200   300   400 Kilometers

0   50   100   150   200 Miles

Yugoslavia, the consequences to regional, and indeed world, stability would be disastrous. Should Serbian ethnic cleansing touch the Hungarian minority in Serbian Vojvodina, that is precisely what would be at risk. But irredentism is not confined to European cases. Tamils in southern India look across the water to the campaign by Sri

Lankan Tamils to carve an independent state from their island country. And Moscow looks to the protection of Russian minorities in neighboring republics.

9. *The Domino Effect*

During the days of the Indochina War, the so-called *domino theory* was much in vogue. Unfortunately, political geographers and others misdefined this important notion, arguing that it holds that communist insurgency in one country will spill over into neighboring countries and destabilize them too. The domino theory does not apply just to insurgencies, communist or otherwise. What the domino theory means is that destabilization in one state, from whatever cause (environmental, political, military), can lead to the destabilization of adjacent states. Historical geography provides many examples: colonial regimes in Africa were affected this way as the wind of change swept eastward from West Africa, then southward to the Cape. More recently, the domino effect is at work in the former Yugoslavia. Note that Slovenia, in the marginal north, was the first to achieve independence; Croatia followed, and then Bosnia was caught up in the maelstrom. The fear now is that the conflict will jump to Kosovo, then affect neighboring Albania and Macedonia, and possibly Greece and Turkey. Eastern Europe's dominoes are precariously balanced.

10. *Religious Fundamentalism*

I noted earlier that religious fundamentalism is a growing force in global affairs. It may not be totally appropriate to enumerate this factor as contributing to the splintering of nations, since the goals of religious fundamentalists in the countries where they are most active is to transform the government of the entire state into a fundamentalist one, not to break up the nation. But this refers particularly to Islamic fundamentalism, when in fact other fundamentalist forces are also at work. Algeria is facing a Muslim fundamentalist challenge, and Egypt is buffeted by one. But the more consequential drama may be in progress in India, where Hindu fundamentalism has arisen with a vengeance—a vengeance aimed principally, but not exclusively, at the country's Muslim minority. A Hindu fundamentalist political party has gained ground, and the specter arises of India, the world's largest democracy, splintering along religious lines. Such an event would have incalculable consequences, but it would reflect what seems to be a new phase of religious revival and fervor, felt from the voting booths of America and the mosques of Iran to the Orthodox sanctuaries of Russia and the streets of India.

11. *Regional Disparities*

Virtually all national governments want, and promote, economic growth. But when they are successful, a geographic conundrum arises: economic progress is never spread equally across the nation. And this means that economic success can create inequalities among regions—inequalities that can risk the cohesion of the state. In the case of China, the communist regime in Beijing confronts just such a problem. Overall, China's economy is growing at a rapid rate—but that fact obscures the reality, which is that China's progress is led by its southern coastal provinces, mainly Guangdong, where economic growth is as much as ten times what it is in China's stagnant interior. In time, the interests of Guangdong and those of Beijing may diverge to such a point that Guangdong may become ungovernable; and this may happen to other coastal zones as well. In that case, China might fall victim to its own success; far from reabsorbing Taiwan, Beijing may have trouble containing all the parts of its mainland. There is a precedent for this: Singapore's split from Malaysia.

CONCLUSION

These geographic forces and factors among many others, are at work to affect the world's course toward a new order. Over the past several years we have heard much about this coming new world order, but in fact there is nothing new about what is happening: some political geographers see parallels between current developments and the period following the signing of the Peace of Westphalia in 1648, at the end of the Thirty Years' War in Europe. The new world order that arose after 1648 had its roots in the free cities of late medieval Europe and in the monarchies of the West. In the words of my colleague Alexander Murphy, the transition then was a case of the medieval giving way to the modern—a modern that was not widely achieved until the late eighteenth and early nineteenth centuries. Out of this process of state formation came the European model that was, at the beginning of the Cold War, the cornerstone of the international system.

Now we have emerged from a Forty Years' War—a Cold War with hot spots—but there will be no Treaty of Moscow. Yet the situation bears some resemblance to that of the seventeenth century: even today, there are states that are essentially feudal, states that are little more than cities with limited hinterlands but wide trading links, states that are still monarchies, states in various stages of democratic evolution, and empires—not to mention the chaos of former Yugoslavia and Transcaucasia.

Out of all this, a new world order must come, but its lineaments are far from clear. In truth, the potential for more disorder has not been greater than it is now. And the collapse of the bipolar world may have created a shortcut to nuclear confrontation. The United States may have emerged as the world's sole superpower, but Russia and three of its neighbors continue to control weaponry that could devastate the Earth in minutes.

If a global treaty were signed, it should in some way accommodate the transformation now affecting states in so many parts of the world. Not all the states that are failing today can objectively be defined as states; some were thrown together by foreign powers and held together by despots. If the state is to remain a stable element in the evolving global system, it will have to be redefined—not on the basis of territorial size or population numbers, but on criteria such as democratic institutions, covenants on human rights, guarantees for minorities, freedoms of cultural expression. The answer to devolution is supranationalism: when states, small or large, are tightly interwoven and interconnected, they are less likely to act unilaterally.

Unlikely as it is, this would also be the time to reform the United Nations. The Security Council should be realigned on the basis of the dozen world geographic realms that form the consensus of regional geographers. The General Assembly, too, should be restructured, also on a regional basis. Peacekeeping operations should be performed in regional contexts, so that Africans, not Americans or Italians, would bear the brunt in such places as Liberia or Somalia, and the French and the Poles, not the Nepalese or the Bangladeshis, in the Balkans.

Which brings us to a hotly debated issue of the day: the size, role, and future of the armed forces of the United States. All nations, all states have histories that include honorable as well as dishonorable acts, and the United States is no exception. But the world is fortunate that of all the contenders during the twentieth century—the Nazis of Berlin, the despots of Moscow, the imperialists of Tokyo—we Americans emerged as (undoubtedly temporary) custodians of global supremacy. It is a responsibility that requires strength as well as restraint—strength not only to confront adversaries, but also to facilitate humanitarian actions. That will be necessary, for our world is not only politically transforming but also in environmental transition. Social upheavals as well as natural catastrophes will confront us, and with a world population still growing at nearly 1 billion per decade, ever larger numbers will be afflicted. Armed forces trained not only to keep the peace but also to bring relief will play a crucial role; South Florida after Hurricane Andrew and Haiti in advance of President Aristide's return are but two recent examples of such operations. In the process, though, the traditional functions of the armed forces of democratic states should not be diminished. On this changing planet, the stability and security of the United States may be the sole bridge to a new world order.

APPENDICES

Nations in the World, UN Member States, and US Diplomatic Relations, as of November 1, 1994

* Diplomatic relations with the United States
+ Member of United Nations

COUNTRY

Short-form name	Long-form name	Capital
Afghanistan *+	Islamic State of Afghanistan	Kābul
Albania *+	Republic of Albania	Tiranë
Algeria *+	Democratic and Popular Republic of Algeria	Algiers
Andorra *+	Principality of Andorra	Andorra la Vella
Angola *+	Republic of Angola	Luanda
Antigua and Barbuda *+	(no long-form name)	Saint John's
Argentina *+	Argentine Republic	Buenos Aires
Armenia *+	Republic of Armenia	Yerevan
Australia *+	Commonwealth of Australia	Canberra
Austria *+	Republic of Austria	Vienna
Azerbaijan *+	Azerbaijani Republic	Baku
Bahamas, The *+	Commonwealth of The Bahamas	Nassau
Bahrain *+	State of Bahrain	Manama
Bangladesh *+	People's Republic of Bangladesh	Dhaka
Barbados *+	(no long-form name)	Bridgetown
Belarus *+	Republic of Belarus	Minsk
Belgium *+	Kingdom of Belgium	Brussels
Belize *+	(no long-form name)	Belmopan
Benin *+	Republic of Benin	Porto-Novo
Bhutan +	Kingdom of Bhutan	Thimphu
Bolivia *+	Republic of Bolivia	La Paz (administrative) Sucre (legislative/judiciary)
Bosnia and Herzegovina *+	Republic of Bosnia and Herzegovina	Sarajevo
Botswana *+	Republic of Botswana	Gaborone
Brazil *+	Federative Republic of Brazil	Brasília
Brunei *+	Negara Brunei Darussalam	Bandar Seri Begawan
Bulgaria *+	Republic of Bulgaria	Sofia
Burkina *+	Burkina Faso	Ouagadougou
Burma *+	Union of Burma	Rangoon

Short-form name	Long-form name	Capital
Burundi *+	Republic of Burundi	Bujumbura
Cambodia *+	Kingdom of Cambodia	Phnom Penh
Cameroon *+	Republic of Cameroon	Yaoundé
Canada *+	(no long-form name)	Ottawa
Cape Verde *+	Republic of Cape Verde	Praia
Central African Republic *+	Central African Republic	Bangui
Chad *+	Republic of Chad	N'Djamena
Chile *+	Republic of Chile	Santiago
China *+	People's Republic of China	Beijing
Colombia *+	Republic of Colombia	Bogotá
Comoros *+	Federal Islamic Republic of the Comoros	Moroni
Congo *+	Republic of the Congo	Brazzaville
Costa Rica *+	Republic of Costa Rica	San José
Côte d'Ivoir	(Ivory Coast)Republic of Côte d'Ivoire	Yamoussoukro
Croatia *+	Republic of Croatia	Zagreb
Cuba+	Republic of Cuba	Havana
Cyprus *+	Republic of Cyprus	Nicosia
Czech Republic *+	Czech Republic	Prague
Denmark *+	Kingdom of Denmark	Copenhagen
Djibouti *+	Republic of Djibouti	Djibouti
Dominica *+	Commonwealth of Dominica	Roseau
Dominican Republic *+	Dominican Republic	Santo Domingo
Ecuador *+	Republic of Ecuador	Quito
Egypt *+	Arab Republic of Egypt	Cairo
El Salvador *+	Republic of El Salvador	San Salvador
Equatorial Guinea *+	Republic of Equatorial Guinea	Malabo
Eritrea *+	State of Eritrea	Asmara
Estonia *+	Republic of Estonia	Tallinn
Ethiopia *+	(no long-form name)	Addis Ababa
Fiji *+	Republic of Fiji	Suva
Finland *+	Republic of Finland	Helsinki
France *+	French Republic	Paris
Gabon *+	Gabonese Republic	Libreville
Gambia, The *+	Republic of The Gambia	Banjul
Georgia *+	Republic of Georgia	T'bilisi
Germany *+	Federal Republic of Germany	Berlin
Ghana *+	Republic of Ghana	Accra
Greece *+	Hellenic Republic	Athens
Grenada *+	(no long-form name)	Saint George's
Guatemala *+	Republic of Guatemala	Guatemala

Short-form name	Long-form name	Capital
Guinea *+	Republic of Guinea	Conakry
Guinea-Bissau *+	Republic of Guinea-Bissau	Bissau
Guyana *+	Co-operative Republic of Guyana	Georgetown
Haiti *+	Republic of Haiti	Port-au-Prince
Holy See *	Holy See	Vatican City
Honduras *+	Republic of Honduras	Tegucigalpa
Hungary *+	Republic of Hungary	Budapest
Iceland *+	Republic of Iceland	Reykjavik
India *+	Republic of India	New Delhi
Indonesia *+	Republic of Indonesia	Jakarta
Iran +	Islamic Republic of Iran	Tehrān
Iraq +	Republic of Iraq	Baghdād
Ireland *+	(no long-form name)	Dublin
Israel *+	State of Israel	See note 1
Italy *+	Italian Republic	Rome
Jamaica *+	(no long-form name)	Kingston
Japan *+	(no long-form name)	Tokyo
Jordan *+	Hashemite Kingdom of Jordan	Amman
Kazakhstan *+	Republic of Kazakhstan	Almaty
Kenya *+	Republic of Kenya	Nairobi
Kiribati *	Republic of Kiribati	Tarawa
Korea, North +	Democratic People's Republic of Korea	P'yŏngyang
Korea, South *+	Republic of Korea	Seoul
Kuwait *+	State of Kuwait	Kuwait
Kyrgyzstan *+	Kyrgyz Republic	Bishkek
Laos *+	Lao People's Democratic Republic	Vientiane
Latvia *+	Republic of Latvia	Riga
Lebanon *+	Republic of Lebanon	Beirut
Lesotho *+	Kingdom of Lesotho	Maseru
Liberia *+	Republic of Liberia	Monrovia
Libya *+	Socialist People's Libyan Arab Jamahiriya	Tripoli
Liechtenstein *+	Principality of Liechtenstein	Vaduz
Lithuania *+	Republic of Lithuania	Vilnius
Luxembourg *+	Grand Duchy of Luxembourg	Luxembourg
Macedonia, The Former Yugoslav Republic of +	The Former Yugoslav Republic of Macedonia	Skopje
Madagascar *+	Republic of Madagascar	Antananarivo
Malawi *+	Republic of Malawi	Lilongwe
Malaysia *+	(no long-form name)	Kuala Lumpur
Maldives *+	Republic of Maldives	Male

Short-form name	Long-form name	Capital
Mali *+	Republic of Mali	Bamako
Malta *+	(no long-form name)	Valletta
Marshall Islands *+	Republic of the Marshall Islands	Majuro
Mauritania *+	Islamic Republic of Mauritania	Nouakchott
Mauritius *+	Republic of Mauritius	Port Louis
Mexico *+	United Mexican States	Mexico
Micronesia, Federated States of *+	Federated States of Micronesia	Palikir
Moldova *+	Republic of Moldova	Chişinău
Monaco *+	Principality of Monaco	Monaco
Mongolia *+	(no long-form name)	Ulaanbaatar
Morocco *+	Kingdom of Morocco	Rabat
Mozambique *+	Republic of Mozambique	Maputo
Namibia *+	Republic of Namibia	Windhoek
Nauru *	Republic of Nauru	Yaren District (no capital city)
Nepal *+	Kingdom of Nepal	Kathmandu
Netherlands *+	Kingdom of the Netherlands	Amsterdam The Hague (seat of gov't)
New Zealand *+	(no long-form name)	Wellington
Nicaragua *+	Republic of Nicaragua	Managua
Niger *+	Republic of Niger	Niamey
Nigeria *+	Federal Republic of Nigeria	Abuja
Norway *+	Kingdom of Norway	Oslo
Oman *+	Sultanate of Oman	Muscat
Pakistan *+	Islamic Republic of Pakistan	Islamabad
Palau (see note 2)	Republic of Palau	Koror
Panama *+	Republic of Panama	Panamá
Papua New Guinea *+	Independent State of Papua New Guinea	Port Moresby
Paraguay *+	Republic of Paraguay	Asunción
Peru *+	Republic of Peru	Lima
Philippines *+	Republic of the Philippines	Manila
Poland *+	Republic of Poland	Warsaw
Portugal *+	Portuguese Republic	Lisbon
Qatar *+	State of Qatar	Doha
Romania *+	(no long-form name)	Bucharest
Russia *+	Russian Federation	Moscow
Rwanda *+	Republic of Rwanda	Kigali
Saint Kitts and Nevis *+	Federation of Saint Kitts and Nevis	Basseterre
Saint Lucia *+	(no long-form name)	Castries

Short-form name	Long-form name	Capital
Saint Vincent and the Grenadines *+	(no long-form name)	Kingstown
San Marino *+	Republic of San Marino	San Marino
Sao Tome and Principe *+	Democratic Republic of Sao Tome and Principe	São Tomé
Saudi Arabia*+	Kingdom of Saudi Arabia	Riyadh
Senegal *+	Republic of Senegal	Dakar
Seychelles *+	Republic of Seychelles	Victoria
Sierra Leone *+	Republic of Sierra Leone	Freetown
Singapore *+	Republic of Singapore	Singapore
Slovakia *+	Slovak Republic	Bratislava
Slovenia *+	Republic of Slovenia	Ljubljana
Solomon Islands*+	(no long-form name)	Honiara
Somalia *+	(no long-form name)	Mogadishu
South Africa *+	Republic of South Africa	Pretoria (administrative) Cape Town (legislative) Bloemfontein (judiciary)
Spain *+	Kingdom of Spain	Madrid
Sri Lanka *+	Democratic Socialist Republic of Sri Lanka	Colombo
Sudan *+	Republic of the Sudan	Khartoum
Suriname *+	Republic of Suriname	Paramaribo
Swaziland *+	Kingdom of Swaziland	Mbabane (administrative) Lobamba (legislative)
Sweden *+	Kingdom of Sweden	Stockholm
Switzerland *	Swiss Confederation	Bern
Syria *+	Syrian Arab Republic	Damascus
Tajikistan *+	Republic of Tajikistan	Dushanbe
Tanzania *+	United Republic of Tanzania	Dar es Salaam
Thailand *+	Kingdom of Thailand	Bangkok
Togo *+	Republic of Togo	Lomé
Tonga *	Kingdom of Tonga	Nuku'alofa
Trinidad and Tobago*+	Republic of Trinidad and Tobago	Port-of-Spain
Tunisia *+	Republic of Tunisia	Tūnis
Turkey *+	Republic of Turkey	Ankara
Turkmenistan *+	(no long-form name)	Ashgabat
Tuvalu *	(no long-form name)	Funafuti
Uganda *+	Republic of Uganda	Kampala

Short-form name	Long-form name	Capital
Ukraine *+	(no long-form name)	Kiev
United Arab Emirates *+	United Arab Emirates	Abu Dhabi
United Kingdom *+	United Kingdom of Great Britain and Northern Ireland	London
United States+	United States of America	Washington, DC
Uruguay *+	Oriental Republic of Uruguay	Montevideo
Uzbekistan *+	Republic of Uzbekistan	Tashkent
Vanuatu *+	Republic of Vanuatu	Port-Vila
Venezuela *+	Republic of Venezuela	Caracas
Vietnam +	Socialist Republic of Vietnam	Hanoi
Western Samoa *+	Independent State of Western Samoa	Apia
Yemen *+ note 3	Republic of Yemen	Sanaa
Zaire *+	Republic of Zaire	Kinshasa
Zambia *+	Republic of Zambia	Lusaka
Zimbabwe *+	Republic of Zimbabwe	Harare

Note 1:   In 1950 the Israel Parliament proclaimed Jerusalem as the capital. The United States, like most other countries that have embassies in Israel, maintains its embassy in Tel Aviv.

Note 2:   Relationship of free association with United States pursuant to Compact of Free Association which entered into force October 1, 1994.

Note 3:   The U.S. view is that the Socialist Federal Republic of Yugoslavia has dissolved and no successor state represents its continuation. Serbia and Montenegro have asserted the formation of a joint independent state, but this entity has not been formally recognized as a state by the United States.

Source: Office of The Geographer and Global Issues, U.S. Department of State, Washington, DC.

Area and Demographic Data for the World's States

(Smallest Microstates Omitted)

	Area (1000 sq mi)	Population (millions) 1994	2000	2010	Annual Rate of Natural Incr. (percent)	Doubling Time (years)	Life Expectancy at Birth (years)	Percent Urban Population	1994 Population Density (per sq mi)	1990 Gross Nat'l Product Per Capita (U.S. dollars)
World	**51,510.8**	**5,602.8**	**6,205.7**	**7,111.5**	**1.7**	**41**	**65**	**43**	**109**	**3,790**
Developed Realms	19,579.3	1,163.8	1,190.4	1,236.5	0.5	148	74	73	59	17,900
Developing Realms	31,931.5	4,439.0	5,015.4	5,875.1	2.0	34	63	34	139	810
Europe	**2,261.0**	**579.7**	**586.9**	**591.7**	**0.2**	**338**	**75**	**73**	**256**	**12,990**
Albania	11.1	3.4	3.8	3.9	1.9	36	73	36	307	—
Austria	32.4	7.9	8.0	8.2	0.1	495	76	55	244	19,240
Belarus	80.2	10.3	10.5	11.1	0.3	217	72	67	129	3,110
Belgium	11.8	10.1	10.2	9.7	0.2	347	76	95	854	15,440
Bosnia-Herzegovina	19.7	4.3	4.5	4.4	0.8	90	72	36	218	—
Bulgaria	42.8	8.9	8.8	8.8	0.0		72	68	207	2,210
Croatia	21.8	4.6	4.6	4.8	0.1	1386	72	51	209	—
Czech Republic	30.6	10.4	10.6	10.8	0.1	1386	72	79	340	—
Denmark	16.6	5.2	5.2	5.1	0.1	753	75	85	312	22,090
Estonia	17.4	1.6	1.6	1.7	0.2	365	70	71	91	3,830
Finland	130.1	5.1	5.2	5.0	0.3	224	75	62	39	26,070
France	211.2	57.3	58.8	58.8	0.4	169	77	73	272	19,480
Germany	137.8	80.5	80.2	78.2	-0.1		75	90	584	20,750
Greece	50.9	10.3	10.4	10.4	0.1	990	75	58	203	6,000
Hungary	35.9	10.3	10.2	10.5	-0.2		70	63	287	2,780
Iceland	39.8	0.3	0.3	0.3	1.2	58	78	90	7	21,150
Ireland	27.1	3.6	3.7	3.4	0.6	122	74	56	132	9,550
Italy	116.3	58.1	58.3	56.4	0.1	1386	76	72	499	16,850
Latvia	24.9	2.7	2.7	2.9	0.1	630	70	71	109	3,410
Liechtenstein	0.06	0.1	0.1	0.1	0.6	110	70	—	506	—
Lithuania	25.2	3.8	3.9	4.1	0.4	2.0	72	69	150	2,710
Luxembourg	1.0	0.4	0.4	0.4	0.3	239	74	78	390	28,770
Macedonia	9.9	2.0	2.1	2.2	1.0	70	72	54	201	—
Malta	0.3	0.4	0.4	0.4	0.8	92	76	85	1,221	6,630
Moldova	13.0	4.4	4.7	5.2	0.8	88	69	48	342	2,170
Netherlands	15.9	15.3	15.8	16.6	0.5	147	77	89	965	17,330
Norway	125.2	4.3	4.4	4.5	0.4	193	77	71	34	23,120
Poland	120.7	38.7	39.5	41.3	0.4	187	71	61	320	1,700
Portugal	34.3	10.5	10.6	10.8	0.1	533	74	30	306	4,890
Romania	91.7	23.2	23.4	24.0	0.1	578	70	54	253	1,640
Serbia-Montenegro	26.9	10.1	10.4	10.8	0.5	141	72	47	375	—

	Area (1000 sq mi)	Population (millions)			Annual Rate of Natural Incr. (percent)	Doubling Time (years)	Life Expectancy at Birth (years)	Percent Urban Population	1994 Population Density (per sq mi)	1990 Gross Nat'l Product Per Capita (U.S. dollars)
		1994	2000	2010						
Slovakia	18.8	5.4	5.6	5.9	0.5	154	71	67	287	—
Slovenia	7.8	1.9	2.0	2.1	0.3	267	73	49	248	—
Spain	194.9	38.7	39.1	40.1	0.2	433	77	91	198	10,920
Sweden	173.7	8.7	8.9	8.9	0.3	210	78	83	50	23,860
Switzerland	15.9	6.9	7.0	6.9	0.3	231	77	60	435	32,790
Ukraine	233.1	52.2	52.4	53.3	0.1	1155	71	68	224	2,340
United Kingdom	94.2	58.1	59.0	59.9	0.3	257	76	90	617	16,070
Russia	**6,592.8**	**150.0**	**152.1**	**162.3**	**0.2**	**301**	**69**	**74**	**23**	**1,820**
Armenia	11.5	3.6	4.0	4.5	1.8	40	72	68	315	2,150
Azerbaijan	33.4	7.4	8.3	9.5	2.0	36	70	53	222	1,670
Georgia	26.9	5.6	5.9	6.1	0.9	80	72	56	207	1,640
North America	**7,509.9**	**287.4**	**301.1**	**327.6**	**0.8**	**89**	**75**	**75**	**38**	**21,580**
Canada	3,831.0	27.8	29.1	32.1	0.8	89	77	78	7	20,450
United States	3,678.9	259.6	272.0	295.5	0.8	89	75	75	71	21,700
Japan	**145.7**	**125.2**	**127.6**	**129.4**	**0.3**	**217**	**79**	**77**	**859**	**25,430**
Australia-New Zealand	**3,069.9**	**21.6**	**22.7**	**25.3**	**0.9**	**81**	**76**	**85**	**7**	**16,366**
Australia	2,966.2	18.1	19.0	21.5	0.8	83	76	85	6	17,080
New Zealand	103.7	3.5	3.7	3.8	1.0	71	75	84	34	12,680
Middle America	**1,055.0**	**159.7**	**183.4**	**208.6**	**2.3**	**32**	**68**	**63**	**151**	**2,143**
Bahamas	5.4	0.3	0.3	0.3	1.5	47	73	75	50	11,510
Barbados	0.2	0.3	0.3	0.3	0.7	102	73	32	1,308	6,540
Belize	8.9	0.2	0.3	0.3	3.1	22	70	52	27	1,970
Costa Rica	19.6	3.3	3.8	4.5	2.4	29	77	45	170	1,910
Cuba	44.2	11.1	11.9	12.3	1.1	62	76	73	251	—
Dominican Republic	18.8	7.8	9.0	9.9	2.0	30	68	58	416	820
El Salvador	8.7	5.9	7.0	7.8	2.9	24	65	48	678	1,100
Guadeloupe	0.7	0.4	0.4	0.4	1.4	50	75	48	571	—
Guatemala	42.0	10.3	12.4	15.8	3.1	22	63	39	246	900
Haiti	10.7	6.8	8.1	9.4	2.9	24	55	29	636	370
Honduras	43.3	5.8	7.0	8.7	3.2	22	64	44	134	590
Jamaica	4.2	2.6	2.9	3.1	2.0	35	73	51	621	1,510
Martinique	0.4	0.4	0.4	0.4	1.2	59	78	82	947	—
Mexico	761.6	91.8	105.4	119.5	2.3	30	69	71	121	2,490
Netherlands Antilles	0.4	0.2	0.2	0.2	1.3	55	74	53	492	—

	Area (1000 sq mi)	Population (millions)			Annual Rate of Natural Incr. (percent)	Doubling Time (years)	Life Expectancy at Birth (years)	Percent Urban Population	1994 Population Density (per sq mi)	1990 Gross Nat'l Product Per Capita (U.S. dollars)
		1994	2000	2010						
Nicaragua	50.2	4.4	5.2	6.4	3.1	23	62	57	87	340
Panama	29.8	2.5	2.8	3.2	1.9	37	73	53	85	1,830
Puerto Rico	3.4	3.8	4.1	3.9	1.2	59	74	72	1,120	6,470
Saint Lucia	0.2	0.2	0.2	0.2	1.7	40	72	46	807	1,900
St. Vincent and the Grenadines	0.2	0.1	0.1	0.1	1.6	43	72	21	594	1,610
Trinidad and Tobago	2.0	1.3	1.4	1.5	1.4	50	70	64	649	3,470
Virgin Is.	0.1	0.1	0.1	0.1	1.7	41	76	39	1,024	—
South America	**6,875.0**	**311.5**	**349.2**	**399.4**	**1.9**	**36**	**67**	**74**	**45**	**2,180**
Argentina	1,068.3	33.9	36.5	40.2	1.2	56	70	86	32	2,370
Bolivia	424.2	8.2	9.6	11.3	2.7	26	61	51	19	620
Brazil	3,286.5	156.5	174.9	200.2	1.9	37	65	74	48	2,680
Chile	292.3	14.1	15.6	17.2	1.8	39	74	85	48	1,940
Colombia	439.7	35.6	40.1	45.6	2.0	35	71	68	81	1,240
Ecuador	109.5	10.5	12.1	14.5	2.4	29	67	55	96	960
French Guiana	35.1	0.1	0.1	0.2	2.2	31	74	81	3	—
Guyana	83.0	0.8	0.9	1.0	1.8	39	64	35	10	370
Paraguay	157.1	4.8	5.6	6.9	2.7	25	67	43	30	1,110
Peru	496.2	23.5	26.7	31.0	2.2	32	61	70	47	1,160
Suriname	63.0	0.5	0.5	0.6	2.0	34	70	48	7	3,050
Uruguay	68.0	3.2	3.3	3.5	0.8	83	72	89	47	2,860
Venezuela	352.1	19.8	23.1	27.3	2.5	27	70	84	56	2,610
North Africa/ Southwest Asia	**7,729.2**	**448.1**	**528.1**	**672.1**	**2.7**	**27**	**63**	**49**	**58**	**—**
Afghanistan	250.0	17.7	20.7	34.5	2.6	27	42	18	71	—
Algeria	919.6	27.5	32.5	37.9	2.8	25	66	50	30	2,060
Bahrain	0.3	0.6	0.6	0.8	2.4	29	72	81	1,854	6,910
Cyprus	3.6	0.7	0.8	0.8	1.1	66	76	62	203	8,040
Djibouti	8.9	0.5	0.5	0.7	2.9	24	48	79	52	—
Egypt	386.7	58.4	67.5	81.3	2.4	28	60	45	151	600
Eritrea*	45.4	2.8	3.3	4.3	2.9	24	48	16	62	—
Iran	636.3	63.7	77.6	105.0	3.3	21	64	54	100	2,450
Iraq	167.9	19.6	24.3	34.1	3.7	19	67	73	117	—
Israel	8.0	5.4	5.9	6.9	1.5	45	76	91	674	10,970
Gaza	0.1	0.8	1.0	1.3	4.6	15	—	—	7,587	—
West Bank	2.3	1.7	2.1	2.4	3.6	19	—	—	746	—
Jordan	35.5	3.8	4.6	6.4	3.4	20	71	70	107	1,240
Kazakhstan	1,049.2	17.4	18.9	21.9	1.4	50	69	58	17	2,470
Kuwait	6.9	1.5	1.7	3.2	3.0	23	74	—	212	—

	Area (1000 sq mi)	Population (millions)			Annual Rate of Natural Incr. (percent)	Doubling Time (years)	Life Expectancy at Birth (years)	Percent Urban Population	1994 Population Density (per sq mi)	1990 Gross Nat'l Product Per Capita (U.S. dollars)
		1994	2000	2010						
Kyrgyzstan	76.6	4.7	5.4	6.6	2.2	31	68	38	61	1,550
Lebanon	4.0	3.6	4.1	4.9	2.1	33	68	84	896	—
Libya	679.4	4.8	5.7	7.1	3.0	23	68	83	7	5,410
Morocco	275.1	27.7	32.0	36.0	2.4	29	64	46	101	950
Oman	82.0	1.7	2.1	3.0	3.5	20	66	11	21	5,650
Qatar	4.3	0.5	0.6	0.7	2.5	28	71	90	118	15,860
Saudi Arabia	830.0	17.2	21.1	31.1	3.5	20	65	77	21	7,070
Somalia	246.2	8.8	10.5	13.9	2.9	24	46	24	36	150
Sudan	976.5	28.2	33.9	42.2	3.1	22	53	20	29	400
Syria	71.5	14.8	18.5	25.6	3.8	18	65	50	207	990
Tajikistan	55.3	5.9	7.1	9.1	3.2	22	69	31	106	1,050
Tunisia	63.2	8.8	9.9	11.3	2.1	33	66	53	139	1,420
Turkey	301.4	61.9	70.5	81.2	2.2	32	66	59	205	1,630
Turkmenistan	188.5	4.1	4.8	5.5	2.7	26	65	45	22	1,700
United Arab Emirates	32.3	2.7	3.1	4.9	2.8	25	71	78	82	19,860
Uzbekistan	172.7	22.5	26.4	32.8	2.7	25	69	40	130	1,350
Yemen	203.9	11.1	13.6	19.0	3.5	20	49	25	55	540
Subsaharan Africa	**8,158.0**	**528.8**	**634.2**	**854.2**	**3.1**	**23**	**52**	**26**	**65**	**529**
Angola	481.4	9.4	11.1	14.9	2.8	25	44	26	20	620
Benin	43.5	5.3	6.4	8.9	3.1	23	47	39	122	360
Botswana	231.8	1.4	1.7	2.4	3.1	23	59	24	6	2,040
Burkina Faso	105.9	10.2	12.4	17.0	3.3	21	52	18	96	330
Burundi	10.8	6.2	7.5	10.1	3.2	21	52	5	574	210
Cameroon	183.6	13.5	16.3	23.1	3.2	22	57	42	73	940
Cape Verde Is.	1.6	0.4	0.5	0.7	3.3	21	61	33	269	890
Central African Republic	240.5	3.3	3.9	4.9	2.6	27	47	43	14	390
Chad	495.8	5.5	6.4	7.7	2.5	28	46	30	11	190
Comoros Is.	0.7	0.5	0.7	0.9	3.5	20	56	26	756	480
Congo	132.1	2.5	3.0	3.9	2.9	24	54	41	19	1,010
Equatorial Guinea	10.8	0.4	0.5	0.6	2.6	26	50	28	36	330
Ethiopia	426.4	54.5	67.6	94.0	2.8	25	47	12	128	120
Gabon	103.4	1.2	1.3	1.4	2.5	28	53	43	11	3,220
Gambia	4.4	1.0	1.1	1.6	2.6	27	44	22	217	260
Ghana	92.1	17.0	20.5	26.9	3.2	22	54	32	185	390
Guinea	94.9	8.2	9.5	11.6	2.5	28	42	22	86	480
Guinea-Bissau	14.0	1.0	1.2	1.5	2.0	35	42	27	75	180
Ivory Coast	124.5	13.9	17.2	25.5	3.6	19	54	43	112	730
Kenya	225.0	28.1	34.9	44.8	3.7	19	61	22	125	370
Lesotho	11.7	2.0	2.4	3.1	2.9	24	58	19	170	470
Liberia	43.0	3.0	3.6	5.5	3.2	22	55	44	69	—
Madagascar	226.7	12.7	15.4	21.3	3.2	22	55	23	56	230

	Area (1000 sq mi)	Population (millions) 1994	2000	2010	Annual Rate of Natural Incr. (percent)	Doubling Time (years)	Life Expectancy at Birth (years)	Percent Urban Population	1994 Population Density (per sq mi)	1990 Gross Nat'l Product Per Capita (U.S. dollars)
Malawi	45.8	9.3	11.4	14.9	3.5	20	49	15	204	200
Mali	478.8	9.1	10.8	14.2	3.0	23	45	22	19	270
Mauritania	398.0	2.2	2.6	3.5	2.8	25	48	41	6	500
Mauritius	0.8	1.1	1.2	1.3	1.5	48	69	41	1,407	2,250
Moçambique	302.3	17.5	20.5	26.6	2.7	26	48	23	58	80
Namibia	318.3	1.5	1.9	2.9	3.1	22	60	27	5	—
Niger	489.2	8.9	10.7	15.1	3.2	22	45	15	18	310
Nigeria	356.7	95.6	113.9	152.2	3.0	23	49	16	268	370
Reunion	1.0	0.6	0.7	0.8	1.8	38	71	62	641	—
Rwanda	10.2	8.3	10.1	14.4	3.4	20	50	7	809	310
São Tome and Principe	0.4	0.1	0.2	0.2	2.5	28	66	38	333	380
Senegal	75.8	8.4	9.9	13.1	2.8	25	48	37	111	710
Sierra Leone	27.7	4.7	5.4	7.3	2.6	27	43	30	168	240
South Africa	471.4	43.9	51.4	66.0	2.6	26	64	56	93	2,520
Swaziland	6.7	0.9	1.1	1.5	3.2	22	55	23	131	820
Tanzania	364.9	29.4	36.1	50.2	3.5	20	52	21	81	120
Togo	21.9	4.1	5.1	7.1	3.7	19	55	24	187	410
Uganda	91.1	18.8	23.3	32.5	3.7	19	51	10	206	220
Zaïre	905.6	40.3	48.6	65.6	3.1	22	52	40	45	230
Zambia	290.6	9.0	11.3	15.5	3.8	18	53	49	31	420
Zimbabwe	150.8	11.0	13.2	17.0	3.1	22	60	26	73	640
South Asia	**1,701.1**	**1,204.6**	**1,370.0**	**1,585.5**	**2.2**	**33**	**58**	**24**	**832**	**337**
Bangladesh	55.6	116.8	134.3	165.1	2.4	29	53	14	2,100	200
Bhutan	18.2	0.8	0.9	1.1	2.0	35	47	13	44	190
India	1,237.1	918.6	1,035.7	1,172.1	2.0	34	59	26	743	350
Maldives	0.1	0.2	0.3	0.4	3.4	20	61	28	2,374	440
Nepal	54.4	20.9	24.2	30.2	2.5	28	50	8	383	170
Pakistan	310.4	129.2	154.7	195.1	3.1	23	56	28	416	380
Sri Lanka	25.3	18.2	19.9	21.4	1.5	46	71	22	718	470
Chinese Realm	**4,394.6**	**1,295.5**	**1,398.6**	**1,534.8**	**1.3**	**54**	**70**	**30**	**459**	**—**
China	3,691.5	1,196.2	1,292.5	1,420.3	1.3	53	70	26	324	370
Hong Kong	0.4	5.8	6.1	6.3	0.7	99	78	100	14,572	11,540
Korea, North	46.5	23.1	25.7	28.5	1.9	37	69	64	496	—
Korea, South	38.0	45.2	48.2	51.7	1.1	65	71	74	1,190	5,400
Macau	0.01	0.5	0.5	0.6	1.3	52	79	97	48,669	—
Mongolia	604.3	2.4	2.8	3.5	2.8	25	65	42	4	112
Taiwan	13.9	21.3	22.8	24.0	1.1	62	74	71	1,532	8,815
Southeast Asia	**1,734.4**	**469.2**	**526.7**	**591.9**	**1.9**	**37**	**62**	**29**	**423**	**933**

	Area (1000 sq mi)	Population (millions)			Annual Rate of Natural Incr. (percent)	Doubling Time (years)	Life Expectancy at Birth (years)	Percent Urban Population	1994 Population Density (per sq mi)	1990 Gross Nat'l Product Per Capita (U.S. dollars)
		1994	2000	2010						
Brunei	2.2	0.3	0.3	0.4	2.5	28	71	59	131	—
Cambodia	69.9	9.5	10.8	10.5	2.2	32	49	13	135	200
Indonesia	741.1	190.9	211.6	238.8	1.7	40	61	31	258	560
Laos	91.4	4.7	5.6	7.2	2.9	24	50	16	51	200
Malaysia	127.3	19.7	22.9	27.1	2.5	27	71	35	155	2,340
Myanmar										
(Burma)	261.2	44.1	49.5	57.7	1.9	36	58	24	169	—
Philippines	115.8	66.8	77.2	85.5	2.4	28	65	43	577	730
Singapore	0.2	2.8	3.1	3.2	1.4	51	75	100	14,206	12,310
Thailand	198.1	58.0	63.1	69.2	1.4	48	67	18	293	1,420
Vietnam	127.2	72.3	82.6	92.4	2.2	31	64	20	569	—
Pacific Realm	**212.4**	**6.1**	**7.0**	**8.5**	**2.3**	**30**	**56**	**19**	**44**	**969**
Fiji	7.1	0.8	0.9	0.9	2.0	35	61	39	110	1,770
Fed. States of										
Micronesia	0.3	0.1	0.1	0.1	2.3	31	—	—	—	—
French Polynesia	1.5	0.2	0.2	0.3	2.3	31	69	58	144	—
New Caledonia	7.4	0.2	0.2	0.2	1.8	39	73	59	25	—
Papua-New										
Guinea	178.3	4.0	4.6	5.7	2.3	31	54	13	23	860
Solomon Islands	11.0	0.4	0.5	0.6	3.6	20	61	9	35	580
Vanuatu	5.7	0.2	0.2	0.3	3.1	22	70	18	33	1,060
Western Samoa	1.1	0.2	0.2	0.3	2.8	25	67	21	187	730

Earth's Extremes

Wettest spot	Mt. Waialeale, Hawai'i	Annual rain average, 471 inches (1,196 centimeters); One-year record, Cherrapunji, India: 1,042 inches (2,647 centimeters) in 1861
Driest spot	Atacama Desert, Chile	Rainfall barely measurable
Coldest spot	Vostok, Antarctica	-127°F (-88°C) recorded in 1960
Hottest spot	Al'Aziziyah, Libya	136°F (58°C) recorded in 1922
Northernmost town	Ny Alesund, Spitsbergen, Norway	
Southernmost town	Puerto Williams, Chile	
Highest town	Aucanquilcha, Chile	17,500 feet (5,334 meters)
Lowest town	Villages along the Dead Sea	1,286 feet (392 meters) below sea level
Largest gorge	Grand Canyon, Colorado River, Arizona	217 miles (349 kilometers) long, 4-13 miles (6-21 kilometers) wide, 1 mile (1.6 kilometers) deep
Deepest gorge	Hells Canyon, Snake River, Idaho	7,900 feet (2,408 meters) deep
Strongest surface wind		231 miles (372 kilometers) per hour; recorded in 1934 at Mount Washington, New Hampshire
Greatest tides	Bay of Fundy, Nova Scotia	53 feet (16 meters)
Biggest crater	Chubb Meteor Crater, Canada	2 miles (3.2 kilometers) wide

Source: National Geographic Society.

The Ten Largest Lakes

Rank	Name	Location	Area (square miles)	(square kilometers)	Greatest Depth (feet)	(meters)	Surface Elevation (feet)	(meters)
1.	Caspian Sea	Western Turkestan	143,240	370,992	3,363	1,025	-92	-28
2.	Superior	USA/Canada Border	31,700	82,103	1,333	406	600	183
3.	Victoria	East Africa	26,820	69,464	279	85	3,720	1,134
4.	Huron	USA/Canada Border	23,000	59,570	750	229	576	176
5.	Michigan	USA	22,300	57,757	923	281	579	176
6.	Aral Sea	Turkestan	14,000*	36,260	145	44	99	30
7.	Tanganyika	East Africa	12,350	31,987	4,800	1,463	2,543	775
8.	Baikal	Eastern Russia	12,160	31,494	5,315	1,620	1,493	445
9.	Great Bear	Northwestern Canada	12,028	31,153	1,356	413	512	156
10.	Malawi	Southern Africa	11,150	28,878	2,280	695	1,550	472

Source: National Geographic Society, Cartographic Division

* The Aral "Sea" has shrunk substantially and its level has dropped accordingly as a result of Soviet-era mismanagement. Before its intakes were diverted for irrigated farming, the Aral Sea's area was nearly 25,000 square miles (64,570 square kilometers), making it the world's fourth largest lake. The figures in this table are estimates of its present dimensions.

Note: A lake is defined as a body of water surrounded by land, irrespective of whether the water is salty or fresh. The Caspian and Aral Seas are misnamed; they are lakes.

The Ten Deepest Lakes

Rank	Name	Location	Greatest Depth (feet)	(meters)
1.	Baikal	Eastern Russia	5,315	1,620
2.	Tanganyika	East Africa	4,800	1,463
3.	Caspian Sea	Western Turkestan	3,363	1,025
4.	Issyk-Kul	Turkestan (Kyrgyzstan)	2,303	702
5.	Malawi	Southern Africa	2,280	695
6.	Great Slave	Northwestern Canada	2,015	614
7.	Great Bear	Northwestern Canada	1,356	413
8.	Superior	USA/Canada Border	1,333	406
9.	Titicaca	South American Andes	990	301
10.	Michigan	USA	923	281

The Ten Largest Seas

Rank	Name	Ocean to Which Adjacent	Surface Area (1000 sq mi)	(1000 sq km)
1.	Coral Sea	Pacific	1,850	4,791
2.	Arabian Sea	Indian	1,492	3,863
3.	South China (alt Nan) Sea	Pacific	1,423	3,685
4.	Weddell Sea	Atlantic	1,080	2,796
5.	Caribbean Sea	Atlantic	1,063	2,754
6.	Mediterranean	Atlantic	969	2,510
7.	Tasman	Pacific	900	2,331
8.	Bering Sea	Pacific	873	2,261
9.	Bay of Bengal	Indian	839	2,172
10.	Gulf of Mexico	Atlantic	596	1,543

Source: Victor Showers, *World Facts and Figures*, 3rd ed., John Wiley & Sons, Inc., 1989. Reprinted with permission.

The Ten Largest Islands

Rank	Name	Owned By	Size (square miles)	(square kilometers)
1.	Greenland	Denmark	840,111	2,175,887
2.	New Guinea	Indonesia, Papua New Guinea	306,212	793,089
3.	Borneo	Indonesia, Malaysia, Brunei	280,101	725,462
4.	Madagascar	Madagascar	226,659	587,046
5.	Baffin	Canada	195,928	507,454
6.	Sumatra	Indonesia	165,098	427,604
7.	Honshu	Japan	87,805	227,415
8.	Britain	United Kingdom	84,201	218,081
9.	Victoria	Canada	83,896	217,291
10.	Ellesmere	Canada	75,766	196,234

Sources: National Geographic Society; National Yearbooks.

The Ten Longest Rivers

Rank	Name	Location	Length (miles)	Length (kilometers)
1.	Nile	Africa	4,150	6,679
2.	Amazon	South America	4,000	6,437
3.	Mississippi-Missouri	North America	3,720	5,987
4.	Chang Jiang (Yangtze)	China	3,720	5,987
5.	Yenisei-Angara	Russia	3,650	5,874
6.	Amur-Argun	East Asia	3,600	5,794
7.	Ob-Irtysh	China, Russia	3,370	5,423
8.	Plata-Parana	South America	3,000	4,828
9.	Huang	China	2,900	4,667
10.	Congo	Africa	2,900	4,667

Sources: U.S. National Oceanic and Atmospheric Administration, *Principal Rivers and Lakes of the World* (1992); National Geographic Society.

The Ten Greatest Waterfalls (Volume of Water)

Rank	Waterfall	River and Location	Average Flow (1000 cubic feet per second)	(1000 cubic meters per second)	Height (feet)	(meters)
1.	Khone	Mekong (alt Khong, Lancang, Lantsang, Mekongk, Tien Giang) River, Kampuchea-Laos	410	11,610	70	21
2.	Niagara	Niagara River, Canada-USA	206	5,830	186	57
3.	Grande	Uruguay (alt Uruguai) River, Argentina-Uruguay	159	4,500	75	23
4.	Paulo Alfonso	São Francisco River, Alagoas-Bahia, Brazil	102	2,890	262	80
5.	Urubupunga	Parana River, Mato Grosso do Sul-São Paulo, Brazil	97	2,750	27	9
6.	Iguacu (alt Iguazu; for Iguassu)	Iguacu (alt Iguazu; for Iguassu River, Argentina-Brazil	60	1,700	230	70
7.	Maribondo (alt Marimbondo)	Grande River, Minas Gerais-São Paulo, Brazil	53	1,500	115	35
8.	Churchill (for Grand)	Churchill (for Hamilton) River, Newfoundland, Canada	49	1,390	245	75
9.	Kabalega (for Murchison)	Nile (off Nil) River, Uganda	42	1,200	130	40
10.	Victoria (alt Mosioa-Tunya)	Zambezi (alt Zambesi, Zambeze) River, Zambia-Zimbabwe	38	1,090	304	92

Source: Victor Showers, *World Facts and Figures*, 3rd ed., John Wiley & Sons, Inc., 1989. Reprinted with permission.

Note: The table includes the conventional and other names of the waterfall, given as *alt* for alternate; the average rate of flow; and the height of the greatest individual leap.

The Ten Highest Mountains

Rank	Name	Mountain Range	Location	Height (feet)	Height (meters)
1.	Everest	Himalayas	Nepal-China Border	29,028	8,848
2.	K2	Karakoram Mountains	Kashmir, India	28,250	8,611
3.	Kanchenjunga	Himalayas	Nepal-India Border	28,208	8,598
4.	Lhotse I	Himalayas	Nepal-China Border	27,923	8,511
5.	Makalu I	Himalayas	Nepal-China Border	27,824	8,481
6.	Lhotse II	Himalayas	Nepal-China Border	27,560	8,400
7.	Dhaulagiri	Himalayas	Nepal	26,810	8,172
8.	Manaslu I	Himalayas	Nepal	26,760	8,157
9.	Cho Oyu	Himalayas	Nepal-China Border	26,750	8,153
10.	Nanga Parbat	Himalayas	Kashmir, India	26,660	8,126

Source: National Geographic Society, Cartographic Division.

The Highest Elevations by Continent

Rank	Landmass Mountain	Highest Range	Mountain	Location	Height (feet)	(meters)
1.	Asia	Everest	Himalayas	Nepal-China Border	29,028	8,848
2.	South America	Aconcagua	Andes	Argentina-Chile Border	22,834	6,960
3.	North America	McKinley	Alaska	Alaska	20,320	6,194
4.	Africa	Kilimanjaro		Tanzania	19,340	5,895
5.	Antarctica	Vinson Massif	Ellsworth	Chilean Claim	16,066	4,897
6.	Europe	Blanc	Alps	French-Italian Border	15,771	4,807
7.	Australia	Kosciusko	Great Dividing	New South Wales	7,310	2,228

Source: National Geographic Society, Cartographic Division.

Note: Mount Elbrus, 18,510 feet (5,642 meters), in the Caucasus Mountains is sometimes listed as Europe's tallest peak. By our regional definition, however, Mount Elbrus lies in Asia, not Europe.

The Ten Largest Deserts

Rank	Name	Location	Estimated Size (square miles)	(square kilometers)
1.	Antarctic*	Anarctica	5,100,000	13,209,000
2.	Sahara	North Africa	3,500,000	9,065,000
3.	Great Australian	Australia	1,200,000	3,108,000
4.	Gobi	Inner Asia	500,000	1,295,000
5.	Great Arabian	Arabian Peninsula	450,000	1,165,500
6.	Kalahara	Southern Africa	250,000	647,500
7.	Taklamakan	China	150,000	388,500
8.	Karakum	Turkestan	120,000	310,800
9.	Thar	India/Pakistan	110,000	284,900
10.	Chihuahuan	Mexico/USA	110,000	284,900

* The desert environment is defined by low annual precipitation, not by temperature. On this basis the bulk of the Antarctic continent, which receives an average water equivalent of 4 inches (102 millimeters) of precipitation, constitutes a desert.

The Coldest Cities

Rank	City and Location	Average Temperature °F	°C	Period of Record
1.	Norilsk, Russia in Asia	12.4	-10.9	17 yrs bef 1925
2.	Yakutsk, Russia in Asia	13.8	-10.1	1931-60
3.	Yellowknife, NWT, Canada	22.3	-5.4	1951-80
4.	Ulan-Bator, Mongolia	23.9	-4.5	1936-60
5.	Fairbanks, Alaska, USA	25.9	-3.4	1951-80
6.	Surgut, Russia in Asia	26.4	-3.1	1884-1960
7.	Chita, Russia in Asia	27.1	-2.7	1890-1919, 1924-56
8.	Nizhnevartovsk, Russia in Asia	27.3	-2.6	1932-60
9.	Hailar, Inner Mongolia, China	27.7	-2.4	1909-42, 1950-53
10.	Bratsk, Russia in Asia	28.0	-2.2	1957-62
11.	Ulan-Ude, Russia in Asia	28.9	-1.7	66 yrs bef 1961
12.	Angarsk, Russia in Asia	29.7	-1.3	1951-60
13.	Whitehorse, Yukon Territory, Canada	29.8	-1.2	1951-80
14.	Irkutsk, Russia in Asia	30.0	-1.1	1881-1960
15.	Yakeshi, Inner Mongolia, China	30.2	-1.0	1914-32,1951-53
16.	Godthab, Greenland	30.7	-0.7	1941-70
16.	Komsomolsk-na-Amure, Russia in Asia	30.7	-0.7	1932-33, 1935-60
18.	Tomsk, Russia in Asia	30.9	-0.6	1881-1960
19.	Kemerovo, Russia in Asia	31.3	-0.4	1933-60

Source: Victor Showers, *World Facts and Figures*, 3rd ed., John Wiley & Sons, Inc., 1989. Reprinted with permission.

The Hottest Cities

Rank	City and Location	Average Temperature °F	°C	Period of Record
1.	Djibouti, Djibouti	86.0	30.0	1912-14, 1939-60
2.	Timbuktu, Mali	84.7	29.3	1951-60
2.	Tirunelveli, Tamil Nadu, India	84.7	29.3	1951-60
2.	Tuticorin, Tamil Nadu, India	84.7	29.3	1951-60
5.	Nellore, Andhra Pradesh, India	84.6	29.2	1931-60
5.	Santa Marta, Colombia	84.6	29.2	1940-42, 1944
7.	Aden, South Yemen	84.0	28.9	1926-60
7.	Madurai, Tamil Nadu, India	84.0	28.9	1931-60
7.	Niamey, Niger	84.0	28.9	1945-60
10.	Hudaydah, North Yemen	83.8	28.8	? yrs bef 1971
10.	Ouagadougou, Burkina Faso	83.8	28.8	25 yrs bef 1960
10.	Thanjavur, Tamil Nadu, India	83.8	28.8	1931-60
10.	Tiruchirapalli, Tamil Nadu,	83.8	28.8	1931-60
14.	Khartoum, Sudan	83.7	28.7	1931-60
14.	Khartoum North, Sudan	83.7	28.7	1931-60
14.	Omdurman, Sudan	83.7	28.7	1931-60
17.	Madras, Tamil Nadu, India	83.5	28.6	1931-60
17.	Port Sudan, Sudan	83.5	28.6	1943-60
19.	Jidda, Hejaz, Saudi Araba	83.3	28.5	1951-60
20.	Tamale, Ghana	83.1	28.4	18 yrs bef 1961

Source: Victor Showers, *World Facts and Figures*, 3rd ed., John Wiley & Sons, In., 1989. Reprinted with permission.

The Wettest Cities

Rank	City and Location	Annual Precipitation (inches)	(millimeters)	Period of Record
1.	Buenaventura, Columbia	265.47	6,743	1910-16, 1956, 1959-61
2.	Monrovia, Liberia	202.01	5,131	1953-56
3.	Pago Pago, American Samoa	196.46	4,990	1932-42, 1945-51
4.	Moulmein, Burma	191.02	4,852	1891-1940
5.	Lae, Papua New Guinea	182.87	4,645	18 yrs bef 1971
6.	Baguio, Luzon I, Philippines	180.04	4,573	1910-38
7.	Sylhet, Bangladesh	175.47	4,457	? yrs bef 1984
8.	Padang, Sumatra I, Indonesia	173.35	4,403	1891-1940
9.	Conakry, Guinea	170.91	4,341	1922-54
10.	Bogor, Java I, Indonesia	166.34	4,225	1933-37
11.	Doula, Cameroon	161.77	4,109	1951-60
12.	Kuching, Sarawak, Malaysia	159.45	4,050	1947-64
13.	Cayenne, French Guiana	147.50	3,744	1951-60
14.	Freetown, Sierra Leone	143.27	3,639	1874-1959
15.	Villavicencio, Colombia	140.20	3,561	6 yrs bef 1951
16.	Ambon, Ambon I, Indonesia	138.98	3,530	1879-1940, 1948-50
17.	Mangalore, Karnataka, India	133.78	3,398	1931-60
18.	Pontianak, Borneo I, Indonesia	131.69	3,345	1879-1950
19.	Bandar Seri Begawan, Brunei	131.0	3,327	5 yrs bef 1967
20.	Colon, Panama	130.75	3,321	1891-1940
21.	Legaspi, Luzon I, Philippines	128.23	3,257	1951-70
22.	Hilo, Hawai'i, USA	128.15	3,255	1951-80

Source: Victor Showers, *World Facts and Figures*, 3rd ed., John Wiley & Sons, Inc., 1989. Reprinted with permission.

The Driest Cities

Rank	City and Location	Annual Precipitation (inches)	(millimeters)	Period of Record
1.	Aswan, Egypt	0.02	0.5	1951-60
2.	Luxor, Egypt	0.03	0.7	1941-60
3.	Arica, Chile	0.04	1.1	? yrs bef 1981
4.	Ica, Peru	0.09	2.3	1954-60
5.	Antofagasta, Chile	0.19	4.9	? yrs bef 1981
6.	Minya, Egypt	0.20	5.1	1945-60
7.	Asyut, Egypt	0.20	5.2	1900-34, 1941-47
8.	Callao, Peru	0.47	12	9 yrs bef 1956
9.	Trujillo, Peru	0.54	14	1963-72
10.	Fayyum, Egypt	0.75	19	1928-34, 1941-47
11.	Chimbote, Peru	0.79	20	1963-72
12.	Suez, Egypt	0.87	22	1910-34, 1941-47
13.	Cairo, Egypt	1.04	26	1909-34, 1945-60
14.	Shibin al Kawm, Egypt	1.05	27	1945-47
15.	Giza, Egypt	1.10	28	1909-34
16.	Hulwan, Egypt	1.18	30	1941-60
17.	Lima, Peru	1.22	31	1931-60
17.	Zagazig, Egypt	1.22	31	23 yrs bef 1948
19.	Ismailia, Egypt	1.49	38	16 yrs bef 1961
20.	Aden, South Yemen	1.61	41	1891-1937, 1941-60
21.	Chiclayo, Peru	1.65	42	1963-72
22.	Piura, Peru	1.69	43	1963-72

Source: Victor Showers, *World Facts and Figures*, 3rd ed., John Wiley & Sons, Inc., 1989. Reprinted with permission.

Outstanding U.S. Geographic Facts

United States (including Alaska & Hawai'i)

Total Area for Fifty States		3,615,211 square miles (9,363,396 square kilometers)
Largest state	Alaska	586,400 square miles (1,518,776 square kilometers)
Smallest state	Rhode Island	1,214 square miles (3,144 square kilometers)
Largest county	San Bernardino County, California	20,125 square miles (52,124 square kilometers)
Northernmost town	Barrow, Alaska	71°18′N
Southernmost city	Hilo, Island of Hawai'i	19°42′N
Southernmost town	Naalehu, Island of Hawai'i	19°4′N (155°35′W)
Easternmost town	Lubec, Maine	66°59′W
Highest point on Atlantic coast	Cadillac Mountain, Mount Desert Island, Maine	1,530 feet (466 meters)
Largest and oldest national park	Yellowstone National Park Wyoming, Montana, Idaho	3,472 square miles (8,992 square kilometers)
Largest national monument	Katmai National Monument, Alaska	4,215 square miles (10,917 square kilometers)
Highest waterfall	Yosemite Falls—total in three sections:	2,425 feet (739 meters)
	Upper Yosemite Fall	1,430 feet (436 meters)
	Cascades in middle section	675 feet (206 meters)
	Lower Yosemite Fall	320 feet (98 meters)
Longest river	Mississippi-Missouri-Red Rock	3,860 miles (6,212 kilometers)
Highest mountain	Mount McKinley, Alaska	20,320 feet (6,194 meters)
Lowest point	Death Valley, California	-282 feet (-86 meters)
Deepest lake	Crater Lake, Oregon	1,932 feet (589 meters) deep
Highest lake	Lake Waiau, Hawai'i	13,020 feet (3,969 meters)
Largest inland bay	Chesapeake Bay	3,237 square miles (8,384 square kilometers)

The Forty-Nine States (including Alaska)
Total area for forty-nine states

3,608,787 square miles
(9,346,758 square kilometers)

The Forty-Eight States
Total area for forty-eight states

3,022,387 square miles
(7,827,982 square kilometers)

Largest state	Texas	267,339 square miles (692,408 square kilometers)
Northernmost town	Penasse, Minnesota	49°22′N
Southernmost city	Key West, Florida	24°33′N
Southernmost mainland town	Florida City, Florida	25°27′N
Highest mountain	Mount Whitney, California	14,495 feet (4,418 meters)

Land and Water

Total area for fifty states	Land: 3,548,974 square miles (9,191,843 square kilometers)	Water: 66,237 square miles (171,554 square kilometers)
Total area for forty-nine states	Land: 3,542,559 square miles (9,175,228 square kilometers)	Water: 66,228 square miles (171,531 square kilometers)
Total area for forty-eight states	Land: 2,971,494 square miles (7,696,169 square kilometers)	Water: 50,893 square miles (131,813 square kilometers)

Source: National Geographic Society

INDEX